NSF-CBMS Regional Conference Series
in Probability and Statistics
Volume 8

ANALYSIS OF LONGITUDINAL AND CLUSTER-CORRELATED DATA

Nan Laird
Harvard University

Institute of Mathematical Statistics
Beachwood, Ohio

American Statistical Association
Alexandria, Virginia

Conference Board of the Mathematical Sciences

Regional Conference Series
in Probability and Statistics

Supported by the
National Science Foundation

The production of the *NSF-CBMS Regional Conference Series in Probability and Statistics* is managed by the Institute of Mathematical Statistics: Patrick Kelly, IMS Production Editor; Julia Norton, IMS Treasurer; and Elyse Gustafson, IMS Executive Director.

Library of Congress Control Number: 2004102584

International Standard Book Number 0-940600-60-9

Printed in the United States of America

Contents

Preface

The analysis of data with outcomes measured repeatedly on each subject has experienced several transforming developments in the last twenty years. This monograph presents a unified treatment of modern methods for longitudinal and/or correlated data that have developed during this period. The basic approach that we take to modeling longitudinal data is to extend familiar univariate regression models to multivariate or correlated outcomes. We deal with linear models for measured data and generalized linear models for binary and count data. We show how methods can accommodate missing outcomes and/or unbalanced designs. Both likelihood and moment methods of estimation are covered, as are random effects approaches to data modeling and parameter estimation.

The monograph assumes that the reader has a solid foundation in statistical inference, linear and generalized linear regression models, and a basic knowledge of multivariate methods. It is appropriate for second year doctoral students or postdoctoral fellows in Statistics/Biostatistics as well as researchers or faculty interested in learning about the field.

Acknowledgments

This monograph grew out of notes written for a course on longitudinal/multivariate data analysis taught by myself and others at the Harvard School of Public Health. I am grateful to many collegues and former students for their comments and contributions to the monograph. I am especially indebted to Andrea Rotnitzky, who wrote drafts of several sections of Chapters 1, 3 and 4. I am also grateful to Stuart Baker, Christl Donnelly, Garrett Fitzmaurice, Joe Hogan, Nick Lange, Stuart Lipsitz, Nick Horton and Jim Ware for collaborations over the years which have contributed greatly to my own work in this area. Roman Torgovitsky provided especially helpful comments on the monograph at a crucial point. Finally, the impetus to turn course notes into a monograph came when Paul Speckman organized a National Science Foundation workshop on longitudinal data at the University of Missouri at Columbia. I am grateful for the NSF support, and to NIGMS for supporting my research in longitudinal data during the last twenty years.

Chapter 1

Introduction to Longitudinal Data Analysis

Longitudinal studies are designed to measure intra-individual change over time. Repeated observations are made on individual subjects, usually at a set of common time points specified by the study protocol. A main objective of longitudinal studies is to relate change over time in individuals to their characteristics (exposure, sex, etc.), or to an experimental condition (drug treatment arm, time since baseline, etc.). In some studies, exposures or experimental conditions may change during the course of the study (as in crossover designs or repeated measures experiments). For example, in the typical crossover design each subject receives every treatment in sequence, with suitable washout periods prior to each treatment. The sequence of treatments is determined a priori by randomization. In repeated measures studies, each subject is measured under a set of pre-specified conditions; differences in the response due to conditions are of primary interest. Outcomes may be measurements, counts, or dichotomous indicators, and we may have multivariate outcomes measured at each of several occasions as well.

In the ideal setting, we will have all subjects measured at the same set of occasions; this greatly facilitates the analysis and interpretation. Some studies may be unbalanced by design, as, for example, when measurements are costly and/or invasive, so only a subset of subjects are measured at all occasions. In other instances it may be very difficult to obtain measurements on all subjects on the same set of occasions. This is especially true when studying human subjects over a long period of follow-up, and when studying clinic populations where illness is a big factor in patient availability. Observations may be mistimed and/or missing, and subjects may drop out or become unavailable for observation.

When the study has a simple, classical design, meaning all subjects are measured on the same set of occasions, and the only covariates which vary over time do so by design, then standard multivariate methods for the analysis of polynomial growth curves, crossover designs, or repeated measures may be used, when outcomes are multivariate normal (see, e.g., Morrison, 1990, and Johnson and Wichern, 1992). When there is substantial missing data, traditional analytic methods can only be applied if one restricts the analysis to the units with complete observations and their validity relies upon the strong and often unrealistic assumptions about the missing data mechanism. Even when these assumptions hold, the complete-case analysis is unsatisfactory, as it discards the information available in the units with incomplete observations.

To some extent, standard univariate regression models and methods can be used to analyze longitudinal data, provided one uses the proper design matrices and takes into account the fact that the observations on individuals are correlated. This approach is the basis of the General Estimating Equations (GEE) approach (Diggle *et al.*, 1994) which we will take up in detail in Chapters 4 and 6. Selection of the design matrix is a key element in all of the methods we discuss, and involves subtleties of model formulation which are crucial in settings with non-standard designs.

Section 1.1 of this chapter will be concerned with types of design matrices for specifying the expected response when the mean is linear in design variables. We will introduce a Linear Model for Correlated Data (LMCD) which can be used to analyze data from any longitudinal study where the mean response can be characterized as linear in design variables. In this case we think of subjects as the basic sampling unit, and response is measured repeatedly on subjects. The model can also be used for the analysis of clustered data; here the sampling unit is the cluster and the repeated observations are individuals in the cluster. This includes traditional clustered survey data, family studies, and nested experimental designs such as those used in animal or teratology studies. Before turning to model development, we consider a few of the data examples which will be presented in later chapters.

Six Cities Study of Air Pollution and Health

The Six Cities Study of Air Pollution and Health (Dockery, 1985) was designed to characterize pulmonary function growth between the ages of six and eighteen and the factors that affect growth. A cohort of 13,379 children born in or after 1967 was enrolled in six communities in the US, designed to have a gradient of air pollution from high to low. Most children were enrolled in the first or second grade and participants were seen annually until high school graduation or loss to follow-up. At each examination, spirometry was performed and a respiratory health questionnaire was completed by a parent or guardian.

Pulmonary function measurements obtained from simple spirometry are widely used as a measure of respiratory health in clinical and epidemiological research. The basic maneuver in simple spirometry is maximal inspiration followed by forced exhalation as rapidly as possible into a closed chamber. Many different measures could be computed from the spirometric curve of volume exhaled versus time; two widely used measures are forced vital capacity (FVC), the total volume of air exhaled, and FEV_1, the volume of air exhaled in the first second of the maneuver.

Because the survey was school-based, children moving out of or into the community during the period were either lost to follow-up or late entrants. Children absent from school during periods of measurement had incomplete records as well. This type of imbalance is commonly found in school- or community-based surveys.

A Clinical Trial in Patients Undergoing an Acute Schizophrenic Episode

Schizophrenia is an incurable disorder characterized by periods of acute psychosis of variable length and intensity. Antipsychotic medication is effective in reducing psychotic behavior in many individuals, but can pose significant adverse side effects. An example of a longitudinal clinical trial is the equivalency trial of a new antipsychotic drug for schizophrenia described in Lapierre *et al.* (1990). This clinical trial was a double-blinded study with randomization between four treatments: three doses (low, medium, and high) of an experimental drug and a control drug with known antipsychotic effects as well as known side effects. Initial studies prior to this double-blinded study suggested that the experimental drug had equivalent antipsychotic activity, with less side effects. The primary objective of this study was the determination of a dose-response relationship for efficacy, tolerability, and safety, and the comparison to the control drug. The study was conducted at 13 clinical centers, and a total of 245 patients were enrolled. The primary efficacy parameter was the Brief Psychiatric Rating Scale (BPRS). This scale measures the extent of a total of 18 observed behaviors, reported behaviors, moods, and feelings, and rates each one on a seven point scale, with a higher number reflecting a worse evaluation. The total BPRS score is the sum of the scores on the 18 items. The stated endpoint was change in BPRS from baseline to six weeks.

Patients were admitted to the hospital for the first four weeks of treatment, and discharged as the clinical condition permitted for the final two weeks. Patients were evaluated at baseline and after one, two, three, four, and six weeks of treatment. One hundred and thirty-four (55%) of the patients completed the study, and eleven additional patients had a six-week evaluation, even though they were technically considered non-completers. The primary reason for discontinuation was a perceived lack of effectiveness of the treatment by the physician; there were also several withdrawals due to side effects. Because patients were hospitalized for the main period of follow-up, there are very few missing observations, except those due to patient removal from protocol. In the event of patient removal, a final measurement was made; then the patient was terminated and no further measurements were made.

AIDS Clinical Trial Data

The AIDS Clinical Trials Group (ACTG) conducts numerous trials designed to evaluate therapies for people infected with Human Immunodeficiency Virus (HIV), which causes AIDS. ACTG Protocol 128 (ACTG, 1993; Brady *et al.*, 1996), hereafter ACTG128, was a multi-center, randomized, double-blinded trial that compared high versus low dose Zidovudine (ZDV) therapy in children born with HIV. Zidovudine, also known as AZT (azidothymidine), inhibits replication of HIV and has been shown to decrease mortality and frequency of opportunistic infection among adults with symptomatic HIV (see ACTG, 1993 and references therein).

One primary endpoint in ACTG128 is neuropsychological development as measured by IQ ratio: children are measured at baseline and every six months for two years (five times total), and some continue to be measured even after being removed from assigned treatment. This practice differs from many trials, where removal from treatment is synonymous with removal from the trial. Reasons for treatment termination prior to two years include undesirable side effects, toxicity, lack of efficacy, and parental decision. Children are not put back on ZDV once removed, but can switch to another treatment regimen. Neuropsychological decline represents advancing disease state, while successful treatment prevents decline.

Muscatine Coronary Risk Factor Study

The Muscatine Coronary Risk Factor Study was a longitudinal study of coronary risk factors in school children (Woolson and Clarke, 1984).

The data set used in our examples contains records on 1,014 children from Muscatine, Iowa, who were 7–9 years old in 1977. Height and weight were measured on each child in three survey years: 1977, 1979, and 1981. Although each child was eligible to participate in all three surveys, data on many children are incomplete. For each survey year, the median weight was calculated for each gender and 1 inch of height. Children with relative weight greater than 110% of the median weight in their respective stratum were classified as obese. This criterion resulted in approximately 20% of the children being described as obese (Woolson and Clarke, 1984).

The repeated binary response of interest is whether the child is described as being obese or not (1 = yes, 0 = no) at each occasion. One of the objectives of this study was to determine trends in obesity in children and the effects of gender and age on risk of obesity in children.

A Comparative trial of a contraceptive drug

Machin *et al.* (1988) discuss problems in the analysis of contraceptive trial data which arise as a result of subject discontinuation or dropout. With modern contraceptive methods, discontinuation due to pregnancy is rare, but it occurs often due to unacceptable side effects, such as irregular bleeding patterns. While time to discontinuation is usually the primary endpoint in these trials, it is also of interest to quantify comparative trends in irregular bleeding patterns, and those patterns which lead to discontinuation of the contraceptive method under study.

A comparative trial of two dosages of depot-medroxyprogesterone acetate (DMPA, 100mg and 50 mg) was conducted by the World Health Organization (1986). Women were administered the drug at three month intervals. The discontinuation rate at the end of the first year of the trial was 40%; more than half of those gave bleeding disturbance as the primary reason for discontinuation.

As part of the trial data collection, women maintained daily diaries recording the presence of any irregular bleeding pattern. The analysis presented by Machin *et al.* (1988) uses a binary indicator of the presence or absence of a specific bleeding disturbance, amenorrhea, in each of the three month periods between injections to study comparative trends in amenorrhea in the two dosage groups, and to study the relationship between trends in amenorrhea and discontinuation.

Familial Aggregation of Lung Cancer in Nonsmokers

Studies of familial aggregation are often designed using the case-control design. Either clinical- or population-based samples of cases and controls (called probands) are selected, then disease and covariate information is obtained on the relatives of the cases and controls. The objective is to study the risk of disease in relatives given the case-control (disease) status of the proband.

A detailed family history study was designed by Schwartz *et al.* (1996) to evaluate family history of lung cancer in a nonsmoker as a risk factor for lung cancer in first-degree relatives. Briefly, cases and controls were selected from participants in a previous study, which included population-based nonsmoking lung cancer cases, ages 40–84 years, diagnosed November 1, 1984 through June 30, 1987. These cases were originally ascertained through the Metropolitan Detroit Cancer Surveillance system of the Karmanos Cancer Institute, a participant in NCI's SEER seer Program. Nonsmokers were identified as individuals reporting never smoking more than 100 cigarettes in their lifetime. Non-smoking controls were matched to the cases on ± 5 years, race, sex and county of residence.

Risk factor data collected for the cases and controls included environmental tobacco smoke exposure, occupational history, history of other lung diseases, and family history. The detailed family history collected on relatives included the above-mentioned risk factor data, smoking history, demographics, and occurrence of cancer and other lung diseases among spouses and first-degree relatives of the probands. Questionnaire data for 2,252 family members of nonsmoking cases and 2,408 family members of 247 nonsmoking controls were obtained. Laird *et al.* (1998) used these data to illustrate the use of multivariate methods in the analysis of family risk data.

1.1 A Linear Model for Correlated Data

The data

Consider a sample of N randomly selected units with n_i measurements of response on each unit, $i = 1, \ldots, N$,

$$\underset{n_i \times 1}{Y_i} = \begin{pmatrix} Y_{i1} \\ \vdots \\ Y_{in_i} \end{pmatrix}$$

where the Y_i are independent vectors and n_i may or may not be the same for all units i. Associated with the jth measurement on the ith unit is a

$p \times 1$ vector of covariates

$$\underset{p\times 1}{X_{ij}} = \begin{pmatrix} X_{ij1} \\ \vdots \\ X_{ijp} \end{pmatrix}$$

and we write

$$\underset{n_i\times p}{X_i} = \begin{pmatrix} X_{i1}^T \\ \vdots \\ X_{in_i}^T \end{pmatrix}.$$

In the design matrix X_i, the rows correspond to the covariates associated with the different times of measurement, and the columns correspond to the different variables. Thus each subject has a vector of outcomes, Y_i, and a matrix of covariates, X_i. In the setting where j indexes occasions of measurements, the covariate X_{ij} may include functions of explanatory variables measured at or prior to time j. There are two main classifications for the type of covariates found in the analysis of correlated data.

a. Classification according to whether the values taken by a covariate associated with measurements on the same unit remain constant or differ.

a.1. We say that the kth covariate, $1 \leq k \leq p$, is between-cluster or time-invariant if for all $i = 1, \ldots, N$,

$$X_{i1k} = X_{i2k} = \cdots = X_{in_ik}$$

Examples include sex and race in a longitudinal study where individuals are the basic sampling unit, and fixed experimental conditions such as treatment assignment in a longitudinal clinical trial. In the cluster sampling setting, an example would be cluster level covariates, e.g., family income, or disease status of the proband in family studies.

a.2. We say that the kth covariate, $1 \leq k \leq p$, is within-cluster or time-varying if for some $i = 1, \ldots, N$, $X_{ijk} \neq X_{ij'k}$ for at least one pair j, j' and $j \neq j'$. Examples include time since baseline, experimental condition in crossover or repeated measure designs, current smoking status, smoking history, current height in a longitudinal study, or individual characteristics in a clustered sample survey. In some cases (pure repeated measures designs, or longitudinal studies with fixed time points), these covariates vary systematically in the same way for each subject so that $X_{ijk} = X_{i'jk}$ for all i and j, for fixed k.

b. Classification according to whether the variable is fixed by design or stochastic

b.1. Covariates which vary systematically over the course of the study but which are fixed by design include treatment group indicators in a crossover design, time since baseline, or individual relationships in a family study.

b.2. Covariates which vary over time and are stochastic include height, current smoking status, or pollution exposure in a longitudinal survey.

Model Formulation

We shall make the following assumptions in formulating a model for relating the response Y_i to the covariates X_i:

(a). $(Y_1, X_1), \ldots, (Y_N, X_N)$ are independently distributed.

(b). Given X_i,

$$E\left(\underset{n_i \times 1}{Y_i}\right) = \underset{n_i \times p}{X_i}\ \underset{p \times 1}{\beta}.$$

(c). Given X_i,

$$\text{cov}\,(Y_i) = \underset{n_i \times n_i}{\Sigma_i},$$

where Σ_i is some known function of the covariates X_i.

Without loss of generality we consider X_i random. Thus, when the jth covariate is fixed by design, X_{ij} takes a fixed value x_{ij} with probability 1. In (b) and (c) and throughout this monograph, in a slight abuse of notation, when X_i is stochastic we eliminate the conditioning covariate X_i in the notation for the conditional mean and covariance of Y_i given X_i. Similarly, when X_i is fixed by design, $E\,(Y_i)$ and $\text{cov}(Y_i)$ refer to the mean and covariance of Y_i taken with respect to the law $f_i\,(y)$ of Y_i. Assumption (a) says that the sample consists of N independently selected units. Assumption (b) says that the conditional mean of the jth outcome of unit i given $X_{i1}, \ldots, X_{in_i}$ is a linear function of X_{ij} only, i.e.,

$$\begin{aligned} E\,(Y_{ij} \mid X_i) &= X_{ij}^T \beta \\ &= \beta_1 X_{ij_1} + \cdots + \beta_p X_{ij_p}. \end{aligned}$$

Assumption (c), cov $(Y_i) = \Sigma_i$, allows for dependencies among measurements on the same unit. The covariance may vary with covariates, e.g., across groups (sex, race), or entries may be functions of time. It may also depend upon i only through n_i. In later sections we will be more explicit about different forms for Σ_i; for now, we leave it very general.

There are important and often overlooked limitations related to the interpretability of the regression model assumed in (b) when covariates are stochastic and time-varying. Regression models for the marginal mean of Y_{ij} as assumed in (b) may be used to answer public-policy related questions for the dependence of the outcome Y_{ij} on X_{ij}. The X_{ij} are defined very generally, and may be functions of explanatory variables $Z_{i1}, \ldots, Z_{ij}$ measured at or prior to time j. When the Z_{ij} are stochastic, and the objective is to infer a causal relationship between changes in Z_{ij} and the outcome Y_{ij} there are important issues regarding the validity and interpretability of the model that we now briefly discuss.

1. Validity: Assumption (b) implies that the conditional expectation of Y_{ij} given the entire covariate process $X_{i1}, \ldots, X_{in_i}$ observed on the ith unit is a function only of the covariate X_{ij}. This assumption may be violated in settings where, conditional on X_{ij}, the current value of the outcome Y_{ij} predicts the subsequent value of the covariate $X_{i(j+1)}$. This is so since the conditional dependence of Y_{ij} and $X_{i(j+1)}$ given X_{ij} may violate the condition

$$E\left(Y_{ij} | X_{ij}, X_{i(j+1)}\right) = E\left(Y_{ij} | X_{ij}\right) \tag{1.1}$$

implied by assumption (b). Equation (1.1) might be violated, for example, in longitudinal studies designed to evaluate the effect of cumulative fat intake at time j on cholesterol level at time j. Letting Z_{ij} denote fat intake of subject i at time j, define $X_{ij} = (Z_{i1}, \ldots, Z_{ij})$, and let Y_{ij} denote the cholesterol level of subject i at time j. With these definitions equation (1.1) might be violated. For example, suppose subjects with high mean cholesterol level at time j who report a high level of fat intake at time j, tend to subsequently reduced their fat intake. If at the same time, subjects with the same level of fat intake, but lower cholesterol at time j tend to report continuously high levels of fat intake, then assumption (b) does not hold.

2. Interpretability: Even if assumption (b) holds, the β parameters may not quantify the parameters of interest in applications. For example, model (b) may correctly specify the dependence of the conditional mean of the cholesterol level measured at the last occasion on the covariates, since at the last occasion, X_{in_i} is a function of the entire fat intake history. Yet, the parameter β may fail to quantify the covariate effect of practical interest when the covariates are stochastic. To simplify our exposition, suppose that all subjects are measured at two time points so that $n_i = 2$, that Z_{ij} is a dichotomous indicator of a high level of fat intake at time j, $j = 1, 2$ and that at baseline, the study population is homogeneous with respect to simultaneous predictors of subsequent fat intake and cholesterol level. One model satisfying assumption (b) might be

$$E\left(Y_{i2}|X_{i2}\right) = \beta_1 + \beta_2 Z_{i2} + \beta_3 Z_{i1} \tag{1.2}$$

where $X_{i2} = (1, Z_{i2}, Z_{i1})$ and $j = 1, 2$. According to model (1.2) the effect on cholesterol at time two of a continuously high fat diet versus a continuously low fat diet is given by

$$E\left(Y_{i2}|Z_{i1} = 1, Z_{i2} = 1\right) - E\left(Y_{i2}|Z_{i1} = 0, Z_{i2} = 0\right) = \beta_2 + \beta_3.$$

The interpretation of $\beta_2 + \beta_3$, however, depends upon implicit model assumptions. If subjects are randomized to high and low fat diets (and adhere to their specified diets), this determines Z_{i1} and Z_{i2}. Hence Z_{i2} is independent of $Y_{i1}|Z_{i1}$, or in general Z_{ij} is independent of

$$(Y_{i1}, \ldots, Y_{i(j-1)}|Z_{i1}, \ldots, Z_{i(j-1)}).$$

In this setting one can assign the desired causal interpretation to $\beta_2 + \beta_3$. In the setting where the predictors are stochastic, one still needs to make independence assumptions to interpret the results causally.

Specifically, $\beta_2 + \beta_3$ measures the causal effect of Z_{i1} and Z_{i2} on Y_{i2} only when *either* the observed cholesterol level at time 1, Y_{i1}, is not a predictor of the observed cholesterol level at time 2, conditional on fat intake history (Z_{i1}, Z_{i2}), i.e.,

$$Y_{i2} \amalg Y_{i1}|Z_{i1}, Z_{i2}, \tag{1.3}$$

or when the observed cholesterol level at time 1 is not a predictor of subsequent fat intake conditional on fat intake at time 1, i.e.,

$$Z_{i2} \amalg Y_{i1}|Z_{i1}. \tag{1.4}$$

Here $X \amalg Y|W$ is used to indicate that X and Y are conditionally independent given W. When (1.4) holds, or in general when

$$Z_{ij} \amalg \left(Y_{i1}, Y_{i2}, \ldots, Y_{i(j-1)}\right)|Z_{i1}, Z_{i2}, .., Z_{i(j-1)} \tag{1.5}$$

is true, the time dependent covariate process Z_{ij} is called an external covariate process (Kalbfleisch and Prentice, 1980). Since (1.3) is not in general a reasonable assumption in a longitudinal setting, assumption (1.5) is the most important in the context of a longitudinal study. An example of such a stochastic predictor might be air pollution levels in a study of lung function growth.

When both (1.3) and (1.4) are false, the parameter β in model (1.2) does not have a causal interpretation even if model (1.2) correctly specifies the dependency of the *actually* observed cholesterol level at time 2 on past fat intake history. This is so because cholesterol level at time 1 (Y_{i1}) is simultaneously a predictor of fat intake at time 2 (Z_{i2}) and an independent risk factor for cholesterol level at time 2 (Y_{i2}). When risk factors for current cholesterol level, such as past cholesterol level, reduce subsequent high fat intake, fat intake specific means of cholesterol level tend to underestimate the true effect of fat intake. For example, the observed cholesterol mean of subjects with an observed continuously high fat intake will be an underestimate of the overall cholesterol mean when all subjects follow a continuously high fat diet, if subjects with high cholesterol level at time 1 tend to reduce their fat intake level more than those with normal cholesterol levels, and if within levels of fat intake, cholesterol levels at time 1 and 2 are correlated.

Study designs

In the context of longitudinal studies we distinguish between *balanced* and *unbalanced* study designs. A design is balanced when all N individuals are to be measured at the same n occasions and it is unbalanced otherwise. Thus in balanced designs, $n_i = n$ so that each Y_i is a $n \times 1$ vector and Σ_i defined in Section 1.2, (c), is a $n \times n$ matrix. We say that a study with a balanced design is *complete* when all n measurements are actually observed on each study participant, and the study is incomplete otherwise. Notice that in an incomplete study with a balanced design the actual number of measurements observed on each subject may be less than or equal n. However, at least conceptually, we could have observed n measurements, $Y_{i1}, \ldots, Y_{in}$, on each subject. In the model formulated in Section 1.2, Y_i is the vector comprised of these, potentially unobserved, outcomes and the matrix Σ_i refers to the covariance of Y_i. If this covariance does not depend on covariates, then it is the same for all the subjects. In this case we eliminate the index i and simply denote it with Σ. We say that the model for the covariance matrix Σ is *unstructured* when its elements are restricted by the condition that Σ be positive definite and symmetric but are otherwise unconstrained. It is often the case that in studies that are initially designed as balanced, measurements are not made at the same exact n occasions, for example due to misstimed events. In this case, even if the study is complete, i.e., if n measurements are actually observed on each subject, a single unstructured covariance matrix is often inappropriate. For example, if measurements can be assumed to have constant variance, a single unstructured covariance would indicate that the correlation between the tth and $(t+1)$th measurement is the same regardless of the time elapsed between the measurements.

In familial aggregation studies or in cluster sampling designs, where the number of subjects in the cluster varies for each cluster, Σ depends on i because its dimension $n_i \times n_i$ is a function of i. Furthermore, it is often the case that Σ also depends on i because the covariance of the measurements made on each cluster varies with covariates. Often, parsimonious covariance models are postulated making the elements of Σ_i depend on a small number of cluster-specific covariates and subject specific covariates, e.g., parent or sibling. In such cases we say that Σ_i is *structured.* We will later consider analysis strategies for both structured and unstructured covariance matrices.

Finally, note that the linear model can be written as

$$\underset{M\times 1}{E\begin{pmatrix} Y_1 \\ \vdots \\ Y_N \end{pmatrix}} = \underset{M\times p}{\begin{pmatrix} X_1 \\ \vdots \\ X_N \end{pmatrix}} \underset{p\times 1}{\beta}$$

or

$$\underset{M\times 1}{E\,(Y)} = \underset{M\times p}{X} \; \underset{p\times 1}{\beta}$$

and

$$\underset{M\times M}{\text{cov}\,(Y)} = \begin{pmatrix} \underset{n_1\times n_1}{\Sigma_1} & 0 & \dots & 0 \\ & \underset{n_2\times n_2}{\Sigma_2} & & 0 \\ & & \ddots & \vdots \\ & & & \Sigma_N \end{pmatrix}, \qquad \text{(Block diagonal matrix)}$$

where $M = \Sigma n_i$. In Chapters 3 through 5 we will additionally assume that, given X_i, Y_i has a multivariate normal distribution. We now consider the formulation of X_i for some specific cases.

1.2 Examples and Special Cases of the Linear Model for Correlated Data

In this section we will show that many "classical" multivariate models are special cases of the LMCD for appropriate choices of X_i, β and Σ_i. We also discuss application in some nonstandard settings.

1. One sample repeated measures.

N subjects are measured repeatedly under n different experimental conditions. The goal is to quantify differences in experimental conditions. The model assumes

$$E(\underset{n\times 1}{Y_i}) = \underset{n\times 1}{\mu} = \begin{pmatrix} 1 & 0 & \cdots & 0 \\ 0 & 1 & \cdots & 0 \\ & & \ddots & \\ & & & 1 \end{pmatrix} \begin{pmatrix} \mu_1 \\ \vdots \\ \mu_n \end{pmatrix}.$$

Here $X_i = I$ (where I denotes the identity matrix) and $\beta = \mu$. In this case, contrasts among the μ_j's are the differences in the outcome means for the different experimental conditions, and as such are of primary interest. Often Σ is left unstructured.

A popular alternative to leaving Σ unstructured is to assume compound symmetry. Under the additional assumption of normal errors, or randomization to the experimental conditions the data can then be analyzed using simple univariate ANOVA methods. We say that Σ has a compound symmetry structure if it can be written as

$$\begin{aligned} \Sigma &= \sigma_\epsilon^2 I + \sigma_\gamma^2 \, 1 \, 1^T \\ &= \begin{pmatrix} \sigma_\epsilon^2 + \sigma_\gamma^2 & \sigma_\gamma^2 & \cdots & \sigma_\gamma^2 \\ & \ddots & & \\ & & \ddots & \\ & & & \sigma_\epsilon^2 + \sigma_\gamma^2 \end{pmatrix}. \end{aligned} \tag{1.6}$$

where 1 denotes an $n \times 1$ vector of one's. Compound symmetry arises by considering the model

$$Y_{ij} = \mu_j + \gamma_i + \epsilon_{ij} \qquad \begin{array}{l} j = 1, \ldots, n \\ i = 1, \ldots, N \end{array}$$

where the γ_i's and the ϵ_{ij}'s are independent of each other, with var $(\gamma_i) = \sigma_\gamma^2$ and var $(\epsilon_{ij}) = \sigma_\epsilon^2$. This implies that with $\epsilon_i = (\epsilon_{i1}, \ldots, \epsilon_{in})^T$

$$Y_i = \mu + \gamma_i \, 1 + \epsilon_i$$

and hence $\text{cov}(Y_i)$ satisfies (1.6).

2. One way multivariate ANOVA (MANOVA)

We assume G treatment groups, and n measurements are obtained for each of N_g subjects in treatment group g, $g = 1, \ldots, G$. Our goal is to test if the mean vector is the same for all G groups. Letting

$$E\left(\underset{n\times 1}{Y_{gi}}\right) = \underset{n\times 1}{\mu_g} \qquad \begin{array}{l} g = 1, \ldots, G \\ i = 1, \ldots, N_g \end{array},$$

we can write

$$E\left(Y_{gi}\right) = \underbrace{\left[\underset{n\times n}{0}, \ldots, 0, \overset{\substack{g\text{th block} \\ \downarrow}}{I}, 0, \ldots, 0\right]}_{n\times(nG)} \underbrace{\begin{bmatrix} \mu_1 \\ \vdots \\ \mu_g \\ \vdots \\ \mu_G \end{bmatrix}}_{(nG)\times 1}.$$

Notice that this model can also be written as

$$\begin{aligned} E\left(Y_{gi}\right) &= (0, \ldots, 1, 0, \ldots, 0) \otimes I \begin{bmatrix} \mu_1 \\ \vdots \\ \mu_G \end{bmatrix} \\ &= a_{gi}^T \otimes I \begin{bmatrix} \mu_1 \\ \vdots \\ \mu_G \end{bmatrix} \end{aligned}$$

where a_{gi} is the indicator vector for the gth group and $\otimes$ is the Kronecker or direct matrix product (Harville, 1999). Here the number of free parameters under the null hypothesis

$$H_0 : \mu_1 = \cdots = \mu_G$$

is equal to n. Under the alternative hypothesis H_1 that at least two groups have unequal mean vectors, i.e.,

$$H_1 : \text{there exist } g_1 \text{ and } g_2 \text{ such that } \mu_{g_1} \neq \mu_{g_2}$$

the number of free parameters p satisfies $n < p \leq nG$. The usual MANOVA assumes $\Sigma_i = \Sigma$ is unstructured and normality of the error terms.

3. One group polynomial growth curve model.

N subjects from the same cohort are observed at the same times denoted by $t_1, \ldots, t_n$; for example, a group of children is observed yearly at ages $6, 7, 8, \ldots, 12$. The linear model for the mean response is expressed as a polynomial in t_j. We might assume a quadratic polynomial

$$E(Y_{ij}) = \beta_0 + \beta_1 t_j + \beta_2 t_j^2,$$

and

$$\underset{n\times 1}{E(Y_i)} = \begin{pmatrix} 1 & t_1 & t_1^2 \\ 1 & t_2 & t_2^2 \\ \vdots & \vdots & \vdots \\ 1 & t_n & t_n^2 \end{pmatrix} \begin{pmatrix} \beta_0 \\ \beta_1 \\ \beta_2 \end{pmatrix} = X\beta.$$

Hence $X_i = X$ and is in general an $n \times q$ matrix of the form

$$X = \begin{pmatrix} 1 & t_1 & t_1^2 & \cdots & t_1^{q-1} \\ 1 & t_2 & t_2^2 & \cdots & t_2^{q-1} \\ \vdots & \vdots & \vdots & \vdots & \vdots \\ 1 & t_n & t_n^2 & \cdots & t_n^{q-1} \end{pmatrix}.$$

One model for Σ_i arises from assuming that each subject has his or her own growth curve with parameters β_i:

$$E(Y_i|\beta_i) = X\,\beta_i \text{ and } \operatorname{var}(Y_i|\beta_i) = \sigma^2 I.$$

Thus each subject has their own vector of growth curve parameters $\beta_i = (\beta_{i0}, \beta_{i1}, \beta_{i2})$, where the β_i are themselves considered random with $E(\beta_i) = \beta$ and var $(\beta_i) = D$. Averaging over the β_i's, we have

$$E(Y_i) \;=\; X\,\beta$$

as before, but now Σ has the variance component type structure

$$\operatorname{var}(Y_i) = \operatorname{var}\,[E(Y_i|\beta_i)] + E\,[\operatorname{var}(Y_i|\beta_i)] = X\,D\,X^T + \sigma^2 I.$$

4. Growth curve with covariates.

As in Example 3, suppose that N subjects are observed at the same n occasions. The model is motivated as follows. Suppose that each subject follows his/her own growth curve indexed by a $q \times 1$ parameter vector β_i. Specifically, for subject i we assume that

$$\underset{n\times 1}{E(Y_i)} = Z\beta_i$$

where the $n \times q$ matrix Z gives the polynomial design on time, e.g., Z is X in the previous example, and the expectation is taken with respect to the law of Y_i, $f_i(y)$. Furthermore, we assume that the covariance of Y_i is the same for all subjects but it has an arbitrary structure, i.e.,

$$\operatorname{cov}(Y_i) = \underset{n \times n}{G}$$

where G is unstructured.

Suppose initially that β_i is a deterministic function of a $q \times p$ time invariant covariate matrix A_i. Specifically, suppose that β_i varies with A_i according to the relationship

$$\beta_i = A_i \beta$$

where β is a $p \times 1$ unknown parameter vector, so that

$$\underset{n \times 1}{E(Y_i)} = Z A_i \beta = X_i \beta.$$

The jth row of A_i gives the design for the regression of the jth component of β_i on the covariates. For example, suppose that in a study of growth in children we assume that height (the outcome) is a linear function of age $(g = 2)$ and further that the intercept is gender specific but the slope is the same for both genders. In this case there is a single time invariant covariate, sex, the row dimension of A_i is equal to 2, and $p = 3$ (two intercepts, but only one slope), so that

$$A_i = \begin{bmatrix} 1 & S_i & 0 \\ 0 & 0 & 1 \end{bmatrix}$$

where $S_i = 1$ for boys and $S_i = 0$ for girls, and $\beta = (\beta_1, \beta_2, \beta_3)$. Under this formulation, β_1 represents the girl's intercept, $\beta_1 + \beta_2$ represents the boys intercept and β_3 is the common slope. Letting $X_i = ZA_i$, we can write

$$E(Y_i) = X_i \beta$$

and $\operatorname{cov}(Y_i) = G$ is unstructured and unknown.

The growth curve model has a special form in the case where each element of β_i has a regression on the same $k \times 1$ vector of covariates a_i. That is, $\beta_{i1} = a_i^T \beta^{(1)}$, $\beta_{i2} = a_i^T \beta^{(2)}, \ldots, \beta_{iq} = a_i^T \beta^{(q)}$, where the $\beta^{(l)}$ are the regression parameters for the lth coefficient. Thus $\beta = (\beta^{(1)T}, \beta^{(2)T}, \ldots, \beta^{(q)T})^T$ and $p = kq$. In this case we may write $A_i = a_i^T \otimes I$ where I is $q \times q$ so that $\beta_i = A_i \beta$, and $E(Y_i) = Z(a_i^T \otimes I)\beta$. This is a very special form because $X_i = ZA_i = Z(a_i^T \otimes I) = a_i^T \otimes Z$. We may also write

$$E(Y_i) = X_i\beta = Z\Delta a_i,$$

where $\Delta(q \times k)$ is a matrix of parameters satisfying $\beta = \text{vec}(\Delta^T)$ and $\text{vec}(\Delta^T)$ is a matrix operator which makes a column vector from a matrix by stacking the columns. The result that $X_i\beta$ is equal to $Z\Delta a_i$ follows from the identity $\text{vec}(ABC) = (C^T \otimes A)\text{vec}(B)$ for conformable matrices A, B, and C. The analysis of this highly structured growth model has been discussed by Grizzle and Allen (1969).

The formulation of the growth curve model with covariates is completed by replacing the assumption that β_i is a deterministic function of A_i with the assumption that β_i is random, it fluctuates around $A_i\beta$ and it has covariance that does not depend on A_i. Specifically, we assume that β_i, $i = 1, \ldots, N$, are independent and satisfy

$$E(\beta_i) = A_i\beta$$

and

$$\text{var}(\beta_i) = \underset{q\times q}{D}.$$

where, if A_i is stochastic, the expectation and covariance are taken with respect to the conditional law of β_i given A_i, and if A_i is fixed by design, they are taken with respect to the law $f_{\beta i}(b)$ of β_i. Thus, with X_i and β defined as above we have that

$$E(Y_i) = X_i\beta$$

but now

$$\begin{aligned}\text{var}(Y_i) &= E\{\text{var}(Y_i|\beta_i)\} + \text{var}\{E(Y_i|\beta_i)\} \\ &= G + ZDZ^T\end{aligned}$$

Often the model is formulated by additionally imposing that $G = \sigma^2 I$, because leaving $G_{n\times n}$ unstructured, does not permit also estimating an additional $q \times q$ matrix of covariance parameter D.

Suppose now that the study design is unbalanced, and in addition that outcomes are not completely observed on all units in the sample. We can still assume the same growth curve, but with subject-specific design matrices which allow each Y_{ij} to be measured at a unique t_{ij}:

$$E(\underset{n_i\times 1}{Y_i}) = \underset{n_i\times q}{Z_i}\,\underset{q\times 1}{\beta_i}.$$

Here the design Z_i depends on occasions of measurement for the ith subject, but the β_i vector remains the same as in the balanced and complete setting.

As before, we let

$$E(\beta_i) = \underset{q\times p}{A_i}\ \underset{p\times 1}{\beta}, \qquad \text{var}(Y_i|\beta_i) = \sigma^2 \underset{n_i\times n_i}{I}, \qquad \text{and var } \beta_i = D, \quad \text{then}$$

$$E(Y_i) = Z_i\, A_i\, \beta = X_i\beta$$

and

$$\text{var}(Y_i) = Z_i\, D\, Z_i^T + \sigma^2 I.$$

Here Σ_i depends on occasions of measurement through Z_i.

5. Epidemiological survey of lung function development in children.

Studies have shown that growth in lung function can be expressed as a linear function of current age and height (Hopper *et al.*, 1991; Laird *et al.*, 1992). Letting response be repeated measures of log forced expiratory volume in one second (log FEV_1), covariates include current age and current height. The mean of log FEV_1 is approximately linear in age and height, implying that we may write

$$\underset{n_i\times 1}{E(Y_i)} = \begin{bmatrix} 1 & a_{i1} & h_{i1} \\ \vdots & \vdots & \vdots \\ 1 & a_{in_i} & h_{in_i} \end{bmatrix} \beta,$$

where (a_{ij}, h_{ij}) are age and height for subject i at occasion j. The issue here is that not all subjects enter the study at the same time; hence we have both "longitudinal" and "cross-sectional" information about lung function dependence on age and height. This will be discussed in Section 1.4. The model for Σ_i is complicated if $\text{var}(Y_{ij})$ depends on both age and height.

6. Family studies: Hypertension in families.

Many family studies are designed to quantify correlation among relatives of a measured factor, such as blood pressure. When the goal is to quantify the inherited genetic component of the correlation, residual correlations are computed which have been adjusted for known or suspected correlates. For example, suppose we have N nuclear families, where each family member is measured on blood pressure and a variety of covariates including age, sex, weight/height, physical activity, diet (various measures), smoking, alcohol, etc. Here i indexes family and j indexes family member. We include all known risk factors in the mean model and then study residual correlation. The specification of Σ_i is complex for specific genetic models and depends on the gene sharing among relatives. As a simple example consider an "association" study restricted to families with both parents and three children. If we assume that: (a) the covariance is independent of the covariates, (b) blood pressure variability is similar in males and females but differs between children and adults, (c) the blood pressure correlation between children and adults is the same for all family members, and (d) the correlation between siblings in the same family is the same, then we have the following simple structure for the covariance matrix

$$\begin{array}{c} \\ M \\ F \\ C_1 \\ C_2 \\ C_3 \end{array} \begin{array}{c} \begin{array}{ccccc} M & F & C_1 & C_2 & C_3 \end{array} \\ \begin{bmatrix} \sigma_A^2 & & & & \\ \sigma_{MF} & \sigma_A^2 & & & \\ \sigma_{CP} & \sigma_{CP} & \sigma_C^2 & & \\ \sigma_{CP} & \sigma_{CP} & \sigma_{CC} & \sigma_C^2 & \\ \sigma_{CP} & \sigma_{CP} & \sigma_{CC} & \sigma_{CC} & \sigma_C^2 \end{bmatrix} \end{array},$$

where σ_A^2 and σ_C^2 are adult and child blood pressure variance; σ_{CP} is covariance between parent and child, etc. The components of Σ_i are of primary interest, especially the correlations between two siblings and a parent-child combination. In the absence of environmental causes for correlation, σ_{MF} should be approximately zero.

1.3 Models for the Variance/Covariance Matrix

In many cases, the primary focus of a study is change in mean response, which is modeled by $X_i\beta$; the parameters indexing Σ_i are nuisance, or possibly of secondary interest. In this setting, if all subjects are measured at the same n occasions and the covariance is assumed not to depend on covariates, $\Sigma_i = \Sigma$ may be taken as an arbitrary $n \times n$ positive definite matrix, and, provided n is small relative to N, the entries of Σ can be estimated with reasonable precision. If n is large relative to N, or if the design is inherently unbalanced, i.e., clustered or with subjects measured at arbitrary time points, then it may be desirable to impose some parametric model on Σ_i.

Structure is usually built into covariance matrices in one of two general ways: using serial correlation models, and using random effects models. The models considered in Examples 3 and 4 of Section 1.2 were particular random effects models. In Chapter 5 we consider random effects models in some detail. Here we consider structures for Σ based on serial correlation.

1. *Banded.* Here

$$\Sigma = \sigma^2 \begin{bmatrix} 1 & \rho_1 & \rho_2 & \cdots & \rho_{n-1} \\ & 1 & \rho_1 & \cdots & \rho_{n-2} \\ & & & \ddots & \vdots \\ & & & & 1 \end{bmatrix}.$$

This formulation is often appropriate when measurements are equally spaced, so that

$$\text{corr}\,(Y_{ij}, Y_{ij+k}) = \rho_k \qquad \text{for all } j \text{ and } k.$$

Notice that the formulation also implies constant variance. A special case of a banded covariance is the autoregressive covariance.

2. *Autoregressive.* Here

$$\Sigma = \sigma^2 \begin{bmatrix} 1 & \rho & \rho^2 & \cdots & \rho^{n-1} \\ & 1 & \rho & \cdots & \rho^{n-2} \\ & & & \ddots & \vdots \\ & & & & 1 \end{bmatrix}.$$

This formulation also assumes equal spacing. The autoregressive model can be generalized to accommodate unequal spacing and number of observations as follows: let $(t_{i1}, \ldots, t_{in_i})$ denote the observation times for the ith subject and let Σ_i be the $n_i \times n_i$ covariance matrix. Let var (Y_{ij}) be constant, and assume that

$$\text{corr}\,(Y_{ij}, Y_{ik}) = h\,(|t_{ij} - t_{ik}|),$$

where $h\,(x)$ is a decreasing function of x that takes values between -1 and 1. Since repeated measurements in longitudinal studies are assumed to be positively correlated, a convenient form for $h\,(\cdot)$ is

$$h\,(d_{ijk}) = \exp\{-\alpha d_{ijk}^b\} \tag{1.7}$$

where $d_{ijk} = |t_{ij} - t_{ik}|$, $b = 1$ or 2, and $\alpha > 0$. This presumes that the correlation is one if measurements are made repeatedly at the same time, and goes to zero rapidly if α is large. Neither of these features is especially realistic for the most common attributes of interest that are measured repeatedly in human subjects longitudinal studies.

Diggle (1988) proposed a model which combines the autoregressive with compound symmetry:

$$\Sigma_i = \sigma^2 I + \delta^2 J + \tau^2 \Omega_i$$

where J is an $n_i \times n_i$ matrix of ones, and Ω_i is a correlation matrix with the jkth element given by $h(d_{ijk})$ as in (1.7), so that

$$\text{var}\ (Y_{ij}) = \sigma^2 + \delta^2 + \tau^2$$

and

$$\rho_{ijk} = \text{corr}\ (Y_{ij}, Y_{ik}) = \frac{\delta^2 + \tau^2\, h(d_{ijk})}{\sigma^2 + \delta^2 + \tau^2}.$$

Notice that this implies that

$$\rho_{ijk} \rightarrow \frac{\delta^2 + \tau^2}{\sigma^2 + \delta^2 + \tau^2} \quad \text{as } d_{ijk} \rightarrow 0,$$

and

$$\rho_{ijk} \rightarrow \frac{\delta^2}{\sigma^2 + \delta^2 + \tau^2} \quad \text{as } d_{ijk} \rightarrow \infty.$$

Here σ^2 can be thought of as sampling variability or measurement error; the variance has three components: sampling, subject to subject, and serial. Repeated measurements made at the same occasions are not perfectly correlated unless $\sigma^2 = 0$ (no sampling error), and provided $\delta^2 > 0$ (subject variability), observations on same subject never go to zero, even if widely separated in time.

1.4 Cross-sectional versus Longitudinal Effects

The simple growth curve model of Example 1.2.4 assumes that individuals are measured at the same occasions in time and have the same age at baseline. This is rarely the case in observational studies. Often, subjects enter the study at different ages and have their measurements taken at different points in time. These unbalanced designs provide the opportunity to obtain information on differences due to aging as well as differences in the response variable across cohorts. However, care must be taken in specifying models and interpreting the results of the analysis because the parameters of the model may not reflect the aging effect as intended.

As a simple example of this situation consider a study in which measurements of an outcome of interest are taken at prespecified, equally spaced, time points $t_1, t_2 = t_1 + t, \ldots, t_n = t_1 + (n-1)\,t$, on each of N independent subjects. Suppose first that a single age-cohort is followed-up so that age of entry to the study is the same, say a, for all subjects. The data then consists of independent vectors $(Y_{i1}, \ldots, Y_{in})$, $i = 1, \ldots, N$, where Y_{ij}, $j = 1, \ldots, n$, is the outcome of the ith subject at age $x_j = a + (j-1)t$. Suppose that it is assumed that the mean increases linearly with age, or equivalently with x_j, so that we may write

$$E\left(Y_{ij}\right) = \mu + \Delta x_j, \qquad j = 1, \ldots, n.$$

Then we have the LMCD model

$$E\begin{bmatrix} Y_{i1} \\ Y_{i2} \\ \cdot \\ \cdot \\ Y_{in} \end{bmatrix} = \begin{bmatrix} 1 & x_1 \\ 1 & x_2 \\ \cdot & \cdot \\ \cdot & \cdot \\ 1 & x_n \end{bmatrix} \begin{pmatrix} \mu \\ \Delta \end{pmatrix} \tag{1.8}$$

Clearly, this model implies

$$E\left(Y_{ij} - Y_{i(j-1)}\right) = \Delta, \qquad \text{for } j = 2, \ldots, n. \tag{1.9}$$

Thus, Δ is the mean outcome change between contiguous time points. It is therefore a measure of the effect of aging; the so called "*longitudinal or intra-subject effect.*"

Suppose now that n age-cohorts are followed according to the following design. As before, measurements are taken only at the prespecified time points $t_1, \ldots, t_n$, and age of entry to the study is the same for all subjects. However, in contrast to the previous design, there is now staggered entry, so that age-cohort 1 enters the study at time t_1, age-cohort 2 enters at time t_2, etc. Even though subjects are the same age at entry to the study, they have different ages at times $t_1, \ldots, t_n$ and differing numbers of measurements. Let Y_{ij} and x_j be defined as before. On subject i we then observe only measurements of $Y_{i1}, Y_{i2}, \ldots, Y_{i(n-C_i+1)}$ where C_i denotes the age-cohort number to which subject i belongs. That is, if $C_i = 1$ we observe $Y_{i1}, \ldots, Y_{in}$ while if $C_i = n$ we observe only Y_{i1}. Note that the index j on Y_{ij} now indicates age of measurement on the ith subject and not occasion of measurement. With only one cohort, age of measurement and occasion of measurement are equivalent, but not with staggered cohorts.

Suppose that for analyzing this unbalanced design we fit model (1.8) where now we omit row j if $j > n - C_i + 1$, i.e., if the observation Y_{ij} is not available for subject i. We now show that the model actually fitted may not appropriately reflect the actual data generating process and, as such, the parameter Δ may no longer have the intended interpretation as quantifying the within-subject effect of aging.

Because Y_{ij} is observed only if $j \leq n - C_i + 1$, then when we fit model (1.8) to the observed data, the actual fitted model is

$$E\left(Y_{ij} | C_i \leq n - j + 1\right) = \mu + \Delta x_j, \tag{1.10}$$

The conditional expectation is the mean response at age $a + x_j t$ for all subjects in cohorts $1, 2, \ldots, n - j + 1$. Note that the conditioning means the model is true for all subjects in cohorts $C_i = 1, \ldots, n-j+1$. Equation (1.10) implies the following interpretation for Δ:

$$\begin{aligned} \Delta = & E\left(Y_{ij} | C_i \leq n - j + 1\right) \\ & - E\left(Y_{i(j-1)} | C_i \leq n - j + 2\right) \qquad \text{for all } j, \end{aligned} \tag{1.11}$$

which can be re-expressed as

$$\begin{aligned} \Delta = & E\left(Y_{ij} - Y_{i(j-1)} | C_i \leq n - j + 1\right) \\ & + \left\{E\left(Y_{i(j-1)} | C_i \leq n - j + 1\right) \right. \\ & \left. - E\left(Y_{i(j-1)} | C_i = n - j + 2\right)\right\} \end{aligned} \tag{1.12}$$

This last equation illustrates the likely misspecification of model (1.10). The first term in the right hand side, $E(Y_{ij} - Y_{i(j-1)} | C_i \leq n - j + 1)$, does measure the mean intra-subject change between ages x_{j-1} and x_j in cohorts $1, 2, \ldots, n - j + 1$. However, the second term,

$$E\left(Y_{i(j-1)} | C_i \leq n - j + 1\right) - E\left(Y_{i(j-1)} | C_i = n - j + 2\right),$$

is a contrast in the mean response at the single age x_{j-1} between cohorts $1, 2, \ldots, n - j + 1$ and cohort $n - j + 2$. This latter measures "secular" or cross-sectional change because it compares individuals in different cohorts. Because Δ does not change with j, model (1.10) implies that this linear combination remains the same for all j. This will only be reasonable when there are no cohort effects. Thus, unless the mean response at entry age a is the same in all cohorts, the parameter Δ cannot be interpreted as a measure of pure longitudinal change.

The point of our example is to illustrate that using naive extensions of models for estimating pure longitudinal change with balanced designs, such as in our problem, a model obtained by simply omitting the rows of the design matrix corresponding to unobserved outcomes, may result in: a) misspecified models or, b) correctly specified models whose parameters no longer retain their interpretation under balanced designs.

Researchers have approached the problem of isolating longitudinal effects in various ways. In the example described above one strategy would be to restrict inferences to subjects measured on all occasions that is, to base inference on data from a single cohort. This strategy will not induce loss of information about longitudinal change when $n = 2$, but will clearly be inefficient otherwise as subjects with incomplete (but more than one) measurements carry information about intra-individual effects. In addition, there is the obvious problem that by restricting attention to a single cohort no inference about cross-sectional effects can be made.

A second general strategy applicable also to continuous outcomes and more complex designs than the one considered in our example is to incorporate subject specific effects into the model. Specifically, assume that the outcomes Y_{ij}, $j = 1, \ldots, n_i$, $i = 1, \ldots, N$ are independently normally distributed with common (unknown) variance σ^2 and mean

$$\mu_{ij} = b_i + \Delta\,(X_{ij} - X_{i1}) \tag{1.13}$$

where the b_i's are fixed unknown constants and X_{ij} is the age of subject i when Y_{ij} is measured. Here we allow for the following possibilities:

1. age and calendar time at which the initial outcome measurement Y_{i1} is obtained vary across subjects, so that subjects belong to different age-cohorts;
2. the number of measurements taken on each individual varies, i.e., n_i changes with i;
3. the outcome measurements are taken at unequally spaced time points.

We use capital letters for X_{ij} to stress its stochastic nature under these unbalanced designs. We can extend model (1.13) to allow for non-linear age trends by considering

$$\mu_{ij} = b_i + h\,(X_{ij} - X_{i1}; \Delta) \tag{1.14}$$

where h is a known function of time from baseline $X_{ij} - X_{i1}$ and an unknown parameter Δ satisfying the constraint of being equal to 0 when $X_{ij} = X_{i1}$. In models (1.13) and (1.14) the values of the b_i's vary with subjects. If the aging process is the same across cohorts then the b_i's absorb the cohorts effects and the parameter Δ reflects pure longitudinal trend. A difficulty with model (1.14) is that the number of nuisance parameters increases with N, hence maximum likelihood gives inconsistent results for inferences (Neyman and Scott, 1951).

Inference about a parameter of interest (in our case Δ) when the number of nuisance parameters (in our case the b_is) grows at the same rate as the sample size can sometimes be resolved by a factorization of the likelihood. If two factors can be found such that one carries most or all of the information about the parameter of interest and is exactly or approximately free of the nuisance parameters. In such case, inference is based on the informative component of the likelihood. In the special case in which μ_{ij} is linear in X_{ij}, and $f(Y_{ij})$ depends upon b_i, X_{ij} and Δ only via μ_{ij}, reparameterization via orthogonalization of the covariate space yields a satisfactory factorization for inference about Δ. Specifically, letting $\tau_i = b_i + (\overline{X}_i - X_{i1})\Delta$ and $Z_{ij} = X_{ij} - \overline{X}_i$, where $\overline{X}_i = \sum_j X_{ij}/n_i$, the mean model is re-expressed as $\mu_{ij} = \tau_i + Z_{ij}\Delta$. It can be easily checked that the distribution of $\widehat{\tau}_i = \sum_j Y_{ij}/n_i = \overline{Y}_i$ does not depend on Δ and is a sufficient statistic for τ_i when Δ and σ^2 take arbitrary fixed values. Further, the conditional distribution of $Y = \{Y_{ij}; i = 1,.., N, j = 1,\ldots,n_i\}$ given $\overline{Y} = (\overline{Y}_1,\ldots,\overline{Y}_N)$, $f(Y|\overline{Y};\Delta,\sigma^2)$ say, is free of of the τ_i's and carries all the available information about Δ and σ^2. This approach is indeed identical to conducting inference based on the differences from the individual means $W_{ij} = Y_{ij} - \overline{Y}_i, j = 1,\ldots,n_i - 1$.

In the more general non-linear setting (1.14) an analogous strategy would be to base inference about Δ on the conditional distribution of Y given a sufficient statistic for b_i, say $S_i, i = 1,\ldots,N$. Besides yielding valid asymptotic inference about Δ, this approach has the advantage of not requiring estimation of the b_i's. However, it can be quite inefficient if S_i is informative about Δ. As a final remark, notice that the method does not use the data of subjects with a single observation, since for them, the conditional distribution of Y_i given S_i is constant.

There are two alternative approaches to conduct inference in the presence of a large number of nuisance parameters that we now mention briefly.

One possibility is to regard the b_i's as realizations from a random variable following a parametric distribution. The large number of nuisance parameters is then replaced by the finite number of unknown parameters indexing the distribution of the b_i's. This formulation falls into the class of random effects models. Inference under these models is discussed in detail in Chapter 5.

A second possibility is to assume that the b_i's follow a deterministic function of cohort. For example, if A_i denotes the age of subject i at a fixed time point t_0, then we may assume that $b_i = \beta_0 + \beta_C A_i$. Under the design with staggered entry previously described this assumption implies that

$$E(Y_{ij}|A_i) = \beta_0 + \beta_C A_i + \Delta(X_{ij} - X_{i1}), \tag{1.15}$$

where we write the conditioning on the random variable A_i to emphasize that the model describes the dependence on time since baseline of the outcome mean of subjects belonging to the same cohort. Further, we can use equation (1.15) to model designs in which the baseline measurement is taken at the same calendar time for all subjects but age of entry to the study varies with subjects (and possibly the subsequent times of measurements also differ across subjects). In this case, X_{i1} is equal to A_i and the model becomes

$$E(Y_{ij}|X_i) = \beta_0 + \beta_C X_{i1} + \Delta(X_{ij} - X_{i1}). \tag{1.16}$$

Equation (1.16) defines an LMCD model with covariates X_{i1} and $X_{ij} - X_{i1}$ associated with the Y_{ij}. Notice that the model implies that

$$E(Y_{i1}|X_i) = \beta_0 + \beta_C X_{i1}$$

and

$$E(Y_{ij} - Y_{i1}|X_i) = \Delta(X_{ij} - X_{i1})$$

Thus, β_C reflects the cross-sectional effect of age since it relates how the mean of the baseline measurement changes with age at baseline, and Δ reflects longitudinal effects, since it relates individual changes in age with changes in the outcome.

As an illustration of this method we consider the study by Ware *et al.* (1990) comparing longitudinal and cross-sectional estimates of decline in pulmonary function. Non smoking adults, ages 25–74 on entry, were examined three times at three year intervals between 1974 and 1983. There were 2,454 white subjects with valid measurements at baseline; 1,973 remained in the study at the first follow-up and 1,713 at the final follow-up. Figure 1.1 represents the qualitative results found.

The cross-sectional curve was obtained by applying ordinary univariate regression methods to the first measurement only, Y_{i1}. The longitudinal curve was obtained by fitting a correlated regression model with outcomes equal to the differences $Y_{i3} - Y_{i2}$ and $Y_{i2} - Y_{i1}$. The cross-sectional curve shows a slower rate of decline. The possible reasons for these results include the following:

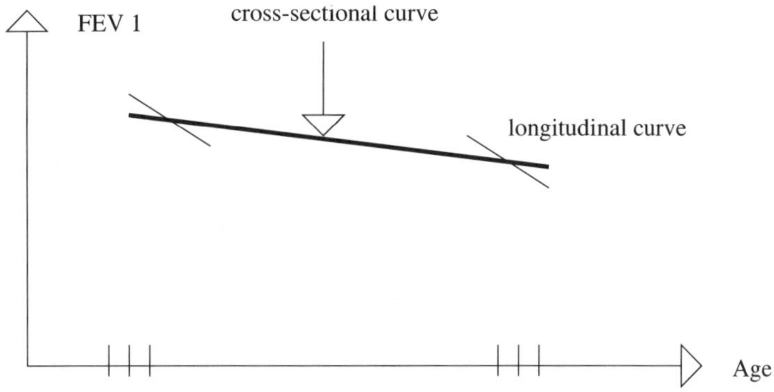

FIGURE 1.1.

- **Cohort effects**:

  Younger cohorts are more prone to respiratory disease because, for example, of higher levels of pollution exposure.

- **Attrition**:

  There might be selection bias, in that at older ages we may only get to see healthier subjects (more subjects with respiratory disease may be dead by age 70 than by age 50).

1.5 Missing Data Issues

The analysis of studies with longitudinal or clustered designs that suffer from non-response poses important methodological challenges. First, we need methods that do not discard units with incomplete measurements but that, instead, efficiently incorporate the observed data on these units. Second, with missing data any analysis method requires important assumptions about the missing data mechanism (MDM) and the usefulness of different methods depends critically on the validity of these assumptions on each specific application. Rubin (1976) and Little and Rubin (1987) provide a classification of the missing data mechanism that is useful for understanding the validity of analysis methods with missing data. To motivate this classification we now provide some examples in the context of the analysis of incomplete correlated outcomes.

An observational longitudinal study with drop-out. The Six Cities Study discussed in Section 1.1 was a longitudinal study of the effects of air pollution on respiratory health. As part of this study, children were followed from age 6 to 18. Lung function was measured with forced expiratory volume (FEV) measurements obtained annually in children in examinations conducted at the schools. Moving in or out of the school district was a predominant reason for late entry or attrition.

An observational longitudinal study with intermittent non-response. As part of the Muscatine Risk Factor Study (Lauer, 1975) weight and height measurements of five age groups of school children in 1977 (ages 5–7, 7–9, 9–11, 11–13, 13–15) were obtained at three examinations conducted every two years. One of the goals of the study was to determine the prevalence, incidence and risk factors for obesity in this cohort. Parental consent prior to each study wave was a requisite for child participation. The study suffered from substantial non-response. For example, of the 471 children age 11-13 in 1977, only 182 participated in the three study waves (Baker, 1995). Other age groups had similarly high non-response rates (Woolson and Clarke, 1984). Some children had intermittent non-response, i.e., they missed an examination but attended a later one. The two main reasons for non-response were: (1) no parental consent was received by the teachers and, (2) the children were absent from school the day of the examination.

A randomized repeated measures study with intermittent non-response. The International Breast Cancer Study Group (IBCSG) trial VI assessed adjuvant chemotherapy in 1461 patients with node-positive breast cancer. Patients were asked to complete a quality of life (QL) questionnaire. Single-item self-assessment scales measured physical well-being, mood, appetite and perceive adjustment/coping. QL was assessed at the beginning of treatment, 2 months after the start of treatment, every three months, and at 1 and 6 months after recurrence. Hurny *et al.* (1996) reported results of the analysis of QL data for the first 18 months. Patients were excluded from the analysis if they could not be classified in any of the nine culture/language categories. In the first 18 months, 1158 (95%) of the 1221 evaluable patients completed at least one questionnaire, 716 (59%) completed at least six, and 415 (34%) completed all seven.

A randomized repeated measures study with drop out and intermittent non-response. The AIDS Clinical Trial Group Protocol 002 (Fischl *et al.* (1990) compared high vs low dose AZT treatment on the health status of AIDS patients. Measurements of CD4 were to be obtained on 520 subjects at baseline and at clinic visits on weeks $t = 8$, 16, 24 and 32. One objective was to compare the evolution in time of mean CD4 counts in the two treatment arms. The study suffered from drop outs and intermittent non-response. Of the 261 patients in the high dose AZT arm, 135 were drop-outs and 22 had intermittent clinic visits.

For the 259 subjects in the low dose AZT arm, these figures were 107 and 25 respectively.

Missing Data Mechanisms. We will now introduce some notation useful for describing the classification of different missing data mechanisms in the context of studies with correlated outcomes. We will assume that for each of N independent units i, $i = 1, \ldots, N$, outcomes Y_{ij}, and a vector of covariates X_{ij}, $j = 1, \ldots, n$, are to be measured. We will let $Y_i = (Y_{i1}, \ldots, Y_{in})^T$ be the complete-data outcome vector, which may not be fully observed and we will let X_i be the $n \times p$ matrix with rows X_{ij}^T. Throughout we will assume that X_i is always observed and that there is no additional information relevant to predicting the outcome Y_i or non-response. Finally, we will let $R_i = (R_{i1}, \ldots, R_{in})$ where $R_{ij} = 1$ if Y_{ij} is observed and $R_{ij} = 0$ otherwise. We will let $Y_{(r)i}$ denote the subvector of Y_i that is observed when $R_i = r$. Thus, $Y_{(R_i)i}$ is the outcome vector actually observed on the ith unit. We will write $f(Y_i|X_i, \beta, \theta)$ for the density of the $n \times 1$ vector Y_i given X_i, where β is the parameter of interest and θ is a nuisance parameter (finite or infinite dimensional) indexing other components of the joint distribution of Y_i given X_i. We will let $P(R_i|Y_i, X_i, \psi)$ denote the conditional probability function of R_i given (Y_i, X_i) which is indexed by the parameter vector ψ. We are now ready to describe the classification of missing data mechanisms (MDM) as originally introduced by Rubin (1976).

Missing completely at random The outcomes are said to be missing completely at random (MCAR) if

$$P(R_i|Y_i, X_i, \psi) = P(R_i|X_i, \psi); \tag{1.17}$$

i.e., the probability of response is conditionally independent of the full outcome vector (observed and unobserved) given the covariates. As an example, in the Six Cities Study the missing lung function measurements would be MCAR if the reasons for missing a lung function test are unrelated to the respiratory health of the child, as would be the case if the child moved out of the school district because of job relocation of the parents. In contrast, the lung function data would not be MCAR if a family decides to relocate into an area with better air quality because of respiratory problems of the child.

Condition (1.17) is equivalent to the conditional independence of the outcome vector Y_i and the response indicator vector R_i given the covariate vector X_i, i.e.,

$$f(Y_i|R_i = r, X_i; \beta, \theta) = f(Y_i|X_i; \beta, \theta) \tag{1.18}$$

Thus, under MCAR the distribution of Y_i is the same in each subpopulation defined by a specific non-response pattern and in the entire target population consisting of the aggregate of units with complete and incomplete responses. Two important practical implications of this fact are the following.

1. If in (1.18) we take r equal to the vector of ones we obtain that the conditional distribution of the outcome vector is the same in the population of units with complete responses as in the target population. Thus, any analysis method that yields valid inference about β in the absence of missing data will also yield valid inference about β if the outcomes are MCAR and the analysis is conducted based on units with complete responses. The latter is usually refer to as a complete case analysis.

2. Equation (1.18) implies that

$$f\left(Y_{(r)i}|R_i = r, X_i; \beta, \theta\right) = f\left(Y_{(r)i}|X_i; \beta, \theta\right) \qquad (1.19)$$

Thus, the conditional distribution of the observed components of Y_i among units with any non-response pattern coincides with the distribution of the same components of Y_i in the target population. In particular, the first and second moments of $Y_{(R_i)i}$ given the covariates are preserved under the MCAR mechanism. That is, if

$$E\left(Y_i|X_i\right) = \underset{n\times p}{X_i}\ \underset{p\times 1}{\beta} \text{ and } \operatorname{cov}\left(Y_i|X_i\right) = \underset{n\times n}{\Sigma_i},$$

then we may write $Y_{(r)i} = I_i Y_i$, where I_i denotes an $n_i \times n$ matrix obtained by removing from the $n \times n$ identity matrix the jth row if $r_{ij} = 0$, $j = 1, \ldots, n$. Hence

$$E\left(Y_{(r)i}|X_i, R_i\right) = E\left(I_i Y_i|X_i\right) = I_i\, E(Y_i|X_i) = \underset{n_i\times n}{I_i}\ \underset{n\times p}{X_i}\ \underset{p\times 1}{\beta} = X_{(r)i}\beta$$

and

$$\operatorname{cov}\left(Y_{(r)i}|X_i, R_i\right) = \underset{n_i\times n}{I_i}\ \underset{n\times n}{\Sigma_i}\ \underset{n\times n_i}{I_i^T} = \Sigma_{(r)i},$$

where $X_{(r)i}$ and $\Sigma_{(r)i}$ denote respectively $I_i X_i$ and $I_i \Sigma_i I_i^T$. Thus, with MCAR outcome data, the appropriate design matrix for the ith subject is simply $X_{(r)i}$ obtained by removing rows of the full data design matrix corresponding to the missing observations. In practice, this has the important implication that standard weighted least squares analyses conducted based on all the available observations of the sampled units yields valid inference about the regression parameter β.

Missing at random The outcomes are said to be missing at random (MAR) if the probability of response is conditionally independent of the unobserved responses given the covariates and the observed responses. That is,

$$P\left(R_i = r|Y_i, X_i; \psi\right) = P\left(R_i = r|Y_{(r)i}, X_i; \psi\right). \tag{1.20}$$

The MAR assumption is attractive because it is less stringent, i.e., it imposes less restrictions on the model for the conditional probability of response given outcomes and covariates, than the MCAR condition. However, an important consequence of the relaxation of assumptions on the model for the missing data mechanism is that the identity (1.19) is no longer guaranteed to be true. This is so because by the Bayes rule,

$$\begin{aligned} & f\left(Y_{(r)i}|R_i = r, X_i; \beta, \theta\right) \\ & \quad = \frac{P\left(R_i = r|Y_{(r)i}, X_i; \psi\right)}{P\left(R_i = r|X_i; \psi\right)} f\left(Y_{(r)i}|X_i; \beta, \theta\right) \end{aligned} \tag{1.21}$$

and the MAR condition does not imply that $P(R_i = r|Y_{(r)i}, X_i; \psi) = P(R_i = r|X_i; \psi)$. This has the important practical implication that complete case analyses as well as least squares analyses based on all available observations no longer yield valid inference about β. However, equation (1.21) implies that the likelihood contribution of unit i can be factored as

$$\begin{aligned} & \prod_r \left\{P\left(R_i = r|X_i; \psi\right) f\left(Y_{(r)i}|R_i = r, X_i; \beta, \theta\right)\right\}^{I(R_i = r)} \\ & \quad = \prod_r \left\{P\left(R_i = r|Y_{(r)i}, X_i; \psi\right)\right\}^{I(R_i = r)} \\ & \qquad \times \prod_r \left\{f\left(Y_{(r)i}|X_i; \beta, \theta\right)\right\}^{I(R_i = r)} \end{aligned} \tag{1.22}$$

where the product ranges over all values taken by r. Thus, if ψ and (β, θ) are variation independent (i.e., the (ψ, β, θ) −parameter space is the Cartesian product of a ψ−parameter space and a (β, θ) −parameter space) and the outcomes are MAR, the likelihood for β is proportional to the likelihood obtained by ignoring the missing data mechanism. When ψ and (β, θ) are variation independent and the outcomes are MAR, the MDM is called *ignorable* (Rubin, 1976). We conclude that under an ignorable MDM, likelihood-based inference does not require the specification of a model for the response probabilities.

The MAR assumption has an easy interpretation when the patterns of non-response are monotone. We say that the non-response patterns are monotone when

$$R_{ij} = 0 \text{ implies } R_{i(j+1)} = 0 \qquad \text{for any } j = 1, \ldots, n-1. \tag{1.23}$$

Monotone patterns arise, for example, in the context of longitudinal studies as a result of data missing solely due to a drop-out process. Under (1.23) it can be easily shown that the MAR condition (1.20) is equivalent to

$$\begin{aligned} &P\left(R_{ij} = 1 | R_{i(j-1)} = 1, Y_{i1}, \ldots, Y_{in}, X_i\right) \\ &\quad = P\left(R_{ij} = 1 | R_{i(j-1)} = 1, Y_{i1}, \ldots, Y_{i(j-1)}, X_i\right) \\ &\qquad\qquad \text{for } j = 1, \ldots, n. \end{aligned} \tag{1.24}$$

Thus, in the context of longitudinal studies with monotone patterns of non-response, MAR means that the probability of dropping out of the study at each time j is conditionally independent of current and future outcomes given the covariates and the observed history of the outcomes up to but not including time j. For example, in the Six Cities Study described previously the outcomes would be missing at random if the decision to move out of the school district was based only on: (a) the air quality (X_i) of the area of residence, (b) the respiratory health history of the child as measured solely by the past recorded FEV test results and (c) other factors unrelated, i.e., (conditionally) statistically independent of, current and future FEV measurements, such as, for example, parental job relocation. In randomized follow-up clinical studies, such as the ACTG trial described previously, the MAR assumption would hold for example if doctors decide to remove patients from the study based on the recorded history of the health outcome variable of interest. Notice, however, that the MAR assumption would not hold if doctors make their decisions based on other health variables predictive of the outcome of interest that are not available for data analysis.

When the longitudinal study also suffers from intermittent nonresponse, i.e., when the monotonicity condition (1.23) does not hold, an MAR process similar in spirit to (1.24) is given by an assumption that non-response depends only on previously observed outcomes, covariates and non-response pattern, i.e.,

$$\begin{aligned} &P\left(R_{ij} = 1 | R_{i1}, \ldots, R_{i(j-1)}, X_i, Y_i\right) \\ &\quad = P\left(R_{ij} = 1 | R_{i1}, \ldots, R_{i(j-1)}, X_i, R_{i1}Y_{i1}, \ldots, R_{i(j-1)}Y_{i(j-1)}\right) \end{aligned} \tag{1.25}$$

When the covariates X_i are successfully obtained from sources other than the study participants, for example from air pollution measurement stations in the communities where the study participants reside as in the

Six Cities Study, Equation (1.25) says that the decision to return to or miss the next study cycle is based solely on the (a) covariates, (b) the actual cycles previously missed and, (c) the outcomes measured at the non-missed prior study cycles. For example, in the context of a randomized clinical follow-up study, assumption (1.25) would hold if X_i recorded treatment arm assignment and baseline patient characteristics and the decision of doctors to temporarily remove patients from the study or to ask patients to return to the study was based solely on measurements of the health outcome of interest that were recorded while the patient was on the study.

Robins and Gill (1997) showed that in contrast to the monotone data case, (1.25) implies MAR but MAR does not imply (1.25). These authors further show that the condition (1.25) imposes restrictions on the law of the observables and it is therefore subject to empirical verification. Unfortunately, as we illustrate below, MAR processes that do not satisfy (1.25) have the unattractive feature that they do not marginalize, that is, unless (1.25) holds,

$$P\left(R_i = r|Y_i\right) = P\left(R_i = r|Y_{(r)i}\right)$$

does not imply

$$P\left(\widetilde{R}_i = \widetilde{r}|\widetilde{Y}_i\right) = P\left(\widetilde{R}_i = \widetilde{r}|\widetilde{Y}_{(\widetilde{r})i}\right)$$

where $\widetilde{R}$, $\widetilde{r}$ and $\widetilde{Y}$ denote the $n-1$ dimensional subvectors of R, r and Y obtained by removal of an arbitrary component. The important practical message of the lack of marginalization is the limitation of the MAR assumption with intermittent non-response when (1.25) does not hold. This is so because any reasonable assumption about the missing data mechanism should not depend on the design, i.e., on the number of observations that are to be potentially taken on each unit. This discussion indicates that when the study suffers from intermittent non-response, practitioners prepared to assume that the data are MAR should routinely test the null hypothesis that condition (1.25) holds and if rejected, they should discount the possibility that the data are MAR.

We now provide a simple example to show that the assumption of MAR does not marginalize when (1.25) is not true. Suppose that $Y_i = (Y_{1i}, Y_{2i})$ is a bivariate vector of binary outcomes and for simplicity suppose that covariates X_i are not relevant. Suppose that $P(R_{1i} = r_1, R_{2i} = r_2|Y_{1i} = y_1, Y_{2i} = y_2)$ is given by the corresponding entries of the following table:

		$(r_1, r_2) = (0,0)$	$(r_1, r_2) = (0,1)$	$(r_1, r_2) = (1,0)$	$(r_1, r_2) = (1,1)$
(y_1, y_2)	$(0,0)$	0.2	0.1	0.3	0.4
	$(0,1)$	0.2	0.2	0.3	0.3
	$(1,0)$	0.2	0.1	0.5	0.2
	$(1,1)$	0.2	0.2	0.5	0.1

Then it can be easily checked that

$$P\left(R_{i1} = 1, R_{i2} = 0 | Y_{1i}, Y_{2i}\right) = P\left(R_{i1} = 1, R_{i2} = 0 | Y_{1i}\right), \tag{1.26}$$

$$P\left(R_{i1} = 0, R_{i2} = 1 | Y_{1i}, Y_{2i}\right) = P\left(R_{i1} = 0, R_{i2} = 1 | Y_{2i}\right), \tag{1.27}$$

and

$$P\left(R_{i1} = 0, R_{i2} = 0 | Y_{1i}, Y_{2i}\right) = P\left(R_{i1} = 0, R_{i2} = 0\right). \tag{1.28}$$

Thus, the data are MAR. However, it follows from the table that

$$P\left(R_{i1} = 0 | Y_{1i} = 0\right) = 0.3P\left(Y_{2i} = 0 | Y_{1i} = 0\right) + 0.4P\left(Y_{2i} = 1 | Y_{1i} = 0\right)$$

and

$$P\left(R_{i1} = 0 | Y_{1i} = 1\right) = 0.3P\left(Y_{2i} = 0 | Y_{1i} = 1\right) + 0.4P\left(Y_{2i} = 1 | Y_{1i} = 1\right),$$

so that $P(R_{i1} = 0)$ depends upon Y_{i1} unless Y_{i2} and Y_{i1} are independent. Hence, (1.26), (1.27) and (1.28) in general do not imply that MAR holds marginally, i.e., in general,

$$P\left(R_{i1} = 0 | Y_{1i}\right) \neq P(R_{i1} = 0),$$

even though $P(R_{i1} = 0, R_{i2} | Y_{i1}, Y_{i2})$ depends only on Y_{i2}.

Non-ignorable. The MDM is non-ignorable when the conditional probability of response depends both on observed and unobserved outcomes i.e., when (1.20) is not true. For example, in the Muscatine Risk Factor Study described previously, the MDM may be non-ignorable because of the possibility of non-response being associated with (unovserved) obesity. In younger children, parents of obese children may be more likely than parents of non-obese children to give consent for participation because of their concern with the health risks associated with obesity. On the other hand, obese preadolescents may be more embarrassed than the non-obese over the examination and more likely to be absent from school on the examination day. In the IBCSG quality of life study, non-ignorable non-response is suspected because patients with worse quality of life may be less likely to want or to be able to fill in a questionnaire. In the ACTG trial 002 non-ignorable non-response is suspected if the possibility exists that doctors' decision to remove patients from study is based not only on the recorded health outcomes from past clinic visits but also on other patient characteristics predictive of prognosis and not available in the data base. Non-response would also be non-ignorable if subjects whose health had markedly deteriorated between visits j and $j+1$ are more likely than others to either miss the clinic visit at week $j+1$ or to never return for a visit after week j.

Important technical and philosophical considerations arise when the MAR assumption is in doubt. Technically, if we are no longer prepared to assume that the outcomes are MAR, we can no longer factorize the individual likelihood contributions as in (1.22). This has the important implication that likelihood-based inference (and indeed any valid inferential procedure) about β under non-ignorable non-response requires the specification of a model for the conditional probability of response. However, as we shall argue in Chapter 9, identification problems arise when fitting richly parameterized models for the non-response probabilities. Philosophically, the values of the parameters indexing the non-response probabilities determine the process of self selection of the non-respondents. Without external information, the data on the respondents only cannot provide any evidence about the self selection process. The identifiability difficulties encountered when fitting non-ignorable non-response models are the mathematical reflection of this paradigm. We will return to these issues in Chapter 9.

In our discussion so far we have assumed that the study design calls for the collection only of data on Y_i and X_i. However, it is often the case that additional time dependent variables V_{ij}, $j = 1, \ldots, n$, not included in the regression model of scientific interest are also recorded and available for data analysis. For example, in the ACTG trial 002 described previously measurements were obtained at each clinic visit not only on CD4 counts (the outcome of interest) but also on other health variables such as white blood cell counts and the occurrence of pneumonia episodes between study cycles. Notice that the regression model of interest does not include as covariates the variables V_i because these variables are in the causal pathway between X_i (treatment) and the outcomes CD4_{ij} and hence their inclusion would lead to over-adjustment. On occasions, we may be prepared to assume that conditional on the covariates and on past measurements of Y_{ij} and V_{ij}, the probability of response at each time j is independent of current and future values of the outcomes. That is,

$$
\begin{aligned}
&P\left(R_{ij}=1|R_{i1},\ldots,R_{i(j-1)},X_i,Y_i,V_i\right)\\
&\quad=P\left(R_{ij}=1|R_{i1},\ldots,R_{i(j-1)},X_i,R_{i1}Y_{i1},R_{i1}V_{i1},\ldots,\right.\\
&\qquad\qquad\left.R_{i(j-1)}Y_{i(j-1)},R_{i(j-1)}V_{i(j-1)}\right).
\end{aligned}
\tag{1.29}
$$

For example, this would be the case in the ACTG trial 002 if a patient misses a study cycle only by indication of the doctor and the doctor's indication is determined solely by the recorded history of Y_{ij} and V_{ij} on the patient.

When, as in the ACTG trial 002, the scientific goal remains the estimation of the parameter β indexing the conditional mean of Y_i on X_i, the analysis presents important methodological challenges if one is not prepared to assume that the more stringent condition (1.25) holds in addition to (1.29). Specifically, as argued in Chapter 9, likelihood based inference about β (and in particular tests comparing the CD4 count means in the two treatment arms) requires the specification of a parametric model for the conditional distribution of the additional variables V_i given the outcomes Y_i and the covariates X_i and can be non-robust to misspecification of this model. In Chapter 9 we discuss likelihood based inference in this setting in more detail.

The methodological difficulties arising when (1.29) holds but (1.25) is not true can be overcome to some extend if the scientific model of interest can be modified to include as covariates the V_i's, because then the outcomes are MAR and likelihood based inference depends only on the correct specification of the model for the conditional law $f(Y_i|X_i,V_i)$. Modification of the scientific model of interest to include the V_is would perhaps be reasonable in certain studies in which the purpose of the analysis is entirely descriptive. For example, consider a survey of lung function in adolescents. We are interested in relating lung function to age and sex to construct norms. But suppose smokers are less likely to participate in the survey. Because smoking is known to be predictive of lung function in adolescents even after stratification on age and sex, an analysis that does not adjust for the differential response rates between smokers and non-smokers would yield biased estimates of the age-sex-specific distribution of lung function in the targeted study population. However, if non-response is independent of lung function after stratifying on age, sex and past smoking status, then (as argued in our discussion of the MAR mechanism) the age-sex-smoking history-specific lung function distribution among respondents is the same as the corresponding strata specific distribution in the targeted population. Thus, in particular, standard regression analyses based on all available observations will result in valid inference about the age-sex-smoking history-specific, but not about the age-sex-specific, lung function means.

Chapter 2

Estimation in the LMCD Assuming Normally Distributed Errors and Complete Data

In this chapter we consider maximum likelihood inference for the parameters of the Linear Model for Correlated Data (LMCD) of Section 1.2 under the additional assumption that the error vectors are normally distributed. To present the basic ideas, we will assume that the data are obtained from a complete study with each $n_i = n$ and we restrict attention to the setting in which Σ_i is assumed not to depend on covariates, so that $\Sigma_i = \Sigma$ is the same for all i. We will consider both maximum likelihood (ML) and restricted maximum likelihood (REML) estimation when Σ is unstructured. Inference under more restrictive parametric models for Σ is briefly discussed at the end of this chapter. Estimation with unbalanced designs and/or missing data will be taken up in Chapter 3.

2.1 ML Estimation of β and Σ

Suppose that in the LMCD we additionally assume that given X_i, Y_i, $i = 1, \ldots, N$, are N copies from a multivariate normal random vector. Then, under this additional assumption, the likelihood of (β, Σ) is given by

$$\mathcal{L}(\beta, \Sigma) = |\Sigma|^{-N/2} \, e^{-\sum_{i=1}^{N}(Y_i - X_i\beta)^T \Sigma^{-1}(Y_i - X_i\beta)/2}.$$

We now characterize the solutions to the score equations

$$\frac{\partial}{\partial\beta}\ln\mathcal{L}(\beta,\,\Sigma)=0 \quad \text{and} \quad \frac{\partial}{\partial\Sigma}\ln\mathcal{L}(\beta,\,\Sigma)=0. \tag{2.1}$$

By convention, for any function $f(M)$ of a $p\times q$ matrix M, we let $\partial f(M)/\partial M$ denote the $p\times q$ matrix with (i,j)th element equal to $\partial f(M)/\partial M_{ij}$. The maximum likelihood estimators of β and Σ, say $\widehat{\beta}_{\text{ML}}$ and $\widehat{\Sigma}_{\text{ML}}$, are solutions of the score equations. The derivatives of $\ln\mathcal{L}(\beta,\,\Sigma)$ are obtained using the following matrix identities (Dwyer, 1967; Harville, 1999). Let x and b be $n\times 1$ vectors and Q be an $n\times n$ symmetric, positive definite matrix, then:

1. $\frac{\partial}{\partial b}(x-b)^T Q^{-1}(x-b)=-2Q^{-1}(x-b)$,
2. $\frac{\partial}{\partial Q^{-1}}\ln\left|Q^{-1}\right|=Q$,
3. $\frac{\partial}{\partial Q^{-1}}(x-b)^T Q^{-1}(x-b)=(x-b)(x-b)^T$.

From

$$\ln\mathcal{L}(\beta,\,\Sigma)=(N/2)\ln|\Sigma^{-1}|-\sum_{i=1}^{N}(Y_i-X_i\beta)^T\,\Sigma^{-1}(Y_i-X_i\beta)/2 \tag{2.2}$$

we use the chain rule, $\partial\ln\mathcal{L}/\partial\beta=\sum_i\left(\partial\mu_i^T/\partial\beta\right)\left(\partial\ln\mathcal{L}/\partial\mu_i\right)$ where $\mu_i=X_i\beta$ to obtain

$$\frac{\partial}{\partial\beta}\ln\mathcal{L}(\beta,\,\Sigma)=-\left\{\sum_{i=1}^{N}X_i^T\Sigma^{-1}(Y_i-X_i\beta)\right\}\Big/\,2$$

and identities 2 and 3 above to obtain

$$\frac{\partial}{\partial\Sigma^{-1}}\ln\mathcal{L}(\beta,\,\Sigma)=\frac{1}{2}\left\{N\Sigma-\sum_{i=1}^{N}(Y_i-X_i\beta)(Y_i-X_i\beta)^T\right\}.$$

Thus, any solution to the score equation $(\widehat{\beta},\,\widehat{\Sigma})$ must satisfy

$$\widehat{\beta}=\left(\sum_{i=1}^{N}X_i^T\widehat{\Sigma}^{-1}X_i\right)^{-1}\left(\sum_{i=1}^{N}X_i^T\widehat{\Sigma}^{-1}Y_i\right) \tag{2.3}$$

and

$$\widehat{\Sigma}=\frac{1}{N}\sum_{i=1}^{N}(Y_i-X_i\widehat{\beta})(Y_i-X_i\widehat{\beta})^T. \tag{2.4}$$

Except in the trivial case of $n = 1$, the solutions to the score equations (2.1) do not have closed form analytic expressions when X_i is arbitrary. There are closed form solutions when X_i and Σ have special forms; estimation for these special cases, and those discussed in the examples of Chapter 1, will be discussed at the end of this chapter.

Equations (2.3) and (2.4) suggest the following iterative algorithm for computing solutions to the score equations. Set $\widehat{\Sigma}_0$ equal to any positive definite matrix, for example $\widehat{\Sigma}_0 = I$. Then for $k \geq 0$ set

$$\widehat{\beta}_k = \left(\sum_{i=1}^{N} X_i^T \widehat{\Sigma}_k^{-1} X_i \right)^{-1} \sum_{i=1}^{N} \left(X_i^T \widehat{\Sigma}_k^{-1} Y_i \right)$$

and

$$\widehat{\Sigma}_{k+1} = \sum_{i=1}^{N} \left(Y_i - X_i \widehat{\beta}_k \right) \left(Y_i - X_i \widehat{\beta}_k \right)^T / N.$$

This algorithm is sometimes referred to as Generalized Least Squares (GLS) (Carroll, Wu and Ruppert, 1988). It can be shown that each iteration of the algorithm increases the likelihood function, i.e., $\mathcal{L}(\widehat{\beta}_k, \widehat{\Sigma}_k) < \mathcal{L}(\widehat{\beta}_{k+1}, \widehat{\Sigma}_{k+1})$, because as we show in Section 3.4, this is also an instance of the EM algorithm. Thus, if the likelihood is bounded, then the sequence of likelihoods, $\mathcal{L}(\widehat{\beta}_k, \widehat{\Sigma}_k)$, $k = 1, 2, \ldots$, converges. The sequence of estimates, $(\widehat{\beta}_k, \widehat{\Sigma}_k)$, $k = 1, 2, \ldots$, however, need not converge but if it does, it converges to a solution of the score equation. Unless the likelihood is concave, this solution need not be equal to the maximum likelihood estimator of (β, Σ). See, for example, Wu (1983) or McLachlan and Krishnan (1996, pp. 92–97). We show in the next chapter that this algorithm can also be derived as an EM algorithm.

2.2 Properties

Even though $\widehat{\beta}$ is a function of $\widehat{\theta}$, it has been shown (Kackar and Harville, 1981) that in the setting where $E(Y_i) = X_i\beta$, $\widehat{\beta}(\widehat{\theta})$is unbiased for β, even in small samples, when the ML (or REML) estimate of θ is used. The basic idea of the proof relies on symmetry of the error distribution (not normality necessarily), and the fact that the variance-covariance estimates are translation invariant, even valued functions of the data. In large samples (see, e.g., Newey and McFadden, 1994, Section 2.4,), $\text{vec}(\widehat{\Sigma}_{\text{ML}})$ is a consistent estimator of $\text{vec}(\Sigma)$ for any error distribution. Here, for any symmetric $q \times q$ matrix A, $\text{vec}(A)$ denotes the $q\,(q+1)\,/2 \times 1$ vector obtained by stacking the columns of the lower diagonal part of A. Furthermore, under additional regularity conditions (Newey and McFadden, 1994, Section 3.2),

$$\left[\begin{array}{c}\sqrt{N}\left\{\widehat{\beta}_{\mathrm{ML}}-\beta\right\}\\ \sqrt{N}\left\{\mathrm{vec}\left(\widehat{\Sigma}_{\mathrm{ML}}\right)-\mathrm{vec}\left(\Sigma\right)\right\}\end{array}\right]\xrightarrow{L} N\left(0,\Gamma\right)$$

where

$$\Gamma=-E\left\{\frac{\partial}{\partial\xi\partial\xi^{T}}\ln\mathcal{L}\left(\beta,\Sigma\right)\right\}^{-1}$$

and $\xi=(\beta^{T},\mathrm{vec}(\Sigma)^{T})^{T}$. Interchanging integration and differentation, the expectation of the mixed derivative matrix

$$\partial\ln\mathcal{L}\left(\beta,\Sigma\right)/\partial\beta^{T}\partial\ \mathrm{vec}\ \left(\Sigma\right)^{T}$$

is equal to zero, so that Γ is a block diagonal matrix with upper and lower squared blocks of dimensions equal to the number of elements of β and $\mathrm{vec}(\Sigma)$ respectively. Thus, $\widehat{\beta}_{\mathrm{ML}}$ and $\widehat{\Sigma}_{\mathrm{ML}}$ are asymptotically independent.

The asymptotic variance of $\widehat{\beta}_{\mathrm{ML}}$ can be easily derived by noticing that $\widehat{\beta}_{\mathrm{ML}}=\widehat{\beta}(\widehat{\Sigma}_{\mathrm{ML}})$. Since $\widehat{\Sigma}_{\mathrm{ML}}$ is a consistent estimator of Σ, then the asymptotic distributions of $\widehat{\beta}_{\mathrm{ML}}$ and $\widehat{\beta}(\Sigma)$ are the same (see Section 4.3). Thus the asymptotic variance of $\widehat{\beta}_{\mathrm{ML}}$ is equal to $E\left\{X_i^{T}\Sigma^{-1}X_i\right\}^{-1}$. This variance can be consistently estimated by

$$\left\{\frac{1}{N}\sum_{i=1}^{N}X_i^{T}\widehat{\Sigma}_{\mathrm{ML}}^{-1}X_i\right\}^{-1}.$$

2.3 (REML) Restricted Maximum Likelihood Estimation

The use of REML as an alternative to maximum likelihood arises in the context of mixed linear models with multiple variance components and was originally proposed by Patterson and Thompson (1971) for analyzing unbalanced block designs.

The motivating factor of REML is to obtain an inferential procedure for Σ that results in estimators and confidence regions that are approximately centered even when the dimension of the mean parameter β is large. The need for such a procedure is justified because inference based on the profile likelihood for Σ (i.e., the maximum likelihood estimator) is misleading when the dimension of β is large relative to the sample size.

As a simple example of this situation and following Barndorff-Nielsen and Cox (1994, Section 3.5) consider inference about the error variance σ^2 in the univariate linear model with normal errors. This corresponds to the case $n = 1$ with $\Sigma = \sigma^2$. The unbiased estimator of σ^2 is given by $\widetilde{\sigma}^2 = SSD/(N-p)$ where SSD is the residual sum of squares and p is the dimension of β. The MLE of $\sigma^2, \widehat{\sigma}^2$, is equal to SSD/N so that $\widehat{\sigma}^2 = \widetilde{\sigma}^2 (N-p)/N$. Suppose now that the dimension of β grows at the same rate as the sample size, that is p/N converges to a positive constant γ as N goes to ∞ (this is, of course, an idealized scenario that nevertheless helps conceptualize the difficulties that arise when the dimension of β is large). Then, with large N, $\widehat{\sigma}^2$ is approximately equal to $(1-\gamma)\widetilde{\sigma}^2$ so that the profile likelihood for σ^2 is sharply peaked at a value distant from the true value of σ^2.

The key idea of REML is to factorize the likelihood for (β, Σ) into two components, one of which is free of the parameter β and which is maximally informative about Σ. To do this, it is convenient to use the following notation: let $Y = (Y_1^T, \ldots, Y_N^T)^T$, $X = (X_1^T, \ldots, X_N^T)^T$, and $\not{\Sigma}$ denote an $Nn \times Nn$ block diagonal matrix with blocks equal to Σ. Then Y is MVN$(X\beta, \not{\Sigma})$. The analysis proceeds by restricting attention to the likelihood based on a vector comprised of the maximum number of linear combinations of the outcomes, say $C_1^T Y, \ldots, C_q^T Y$, such that the $Nn \times 1$ vectors $C_1, \ldots, C_q$ are linearly independent and the distribution of $H^T Y$ does not depend on β, where $H^T = (C_1, C_2, \ldots, C_q)$. Harville (1977) has called the components of $H^T Y$ "error contrasts." The name arises because if $C^T Y$ has a distribution independent of β, then in particular $E(C^T Y) = C^T X\beta$ does not depend on β. But this occurs for $\beta \neq 0$ if and only if

$$C^T X = 0 \tag{2.5}$$

and so $C^T Y$ has mean zero and can be interpreted as an "error."

If, as we shall assume, X is of full rank p, then the subspace of $Nn \times 1$ vectors satisfying (2.5) is of dimension $q = Nn - p$. Thus, $H^T Y$ is a vector of dimension $q \times 1$ with elements $C_j^T Y$, $j = 1, \ldots, q$, where $C_1, \ldots, C_q$ are any subset of linearly independent $Nn \times 1$ vectors satisfying (2.5). Computation of the likelihood based on $H^T Y$ is simplified if we consider the specific error contrast vector determined by H satisfying $H^T H = I$ and $HH^T = A$ where $A = I - X\left(X^T X\right)^{-1} X^T$, (such H always exists because A is symmetric and idempotent). That H determines an error contrast vector can be readily seen from $H^T X = \left(H^T H\right) H^T X = H^T A X = 0$, where the last identity follows because $AX = 0$. There is no loss of generality in this choice, since any other full rank set of error contrasts may be obtained from H^T as $F^T = P^T H^T$, where P^T is an $(Nn - p)$ dimensional square, orthogonal matrix, and it is easily verified that $F^T F = I$ and $FF^T = A$.

Formally, the REML likelihood is defined as the likelihood of $H^T Y$. We will now show that the REML likelihood can also be written as

$$L_{\text{REML}}(\Sigma) = L_{\text{ML}}(\widehat{\beta}, \Sigma) / \left|X^T \Sigma^{-1} X\right|^{1/2},$$

where $L_{\text{ML}}(\widehat{\beta}, \Sigma)$ is the profile likelihood for Σ or alternatively,

$$L_{\text{REML}}(\Sigma) \propto f(Y; \beta, \Sigma) / f(\widehat{\beta}; \beta, \Sigma),$$

where for simplicity, we write $\widehat{\beta}(\Sigma) = \widehat{\beta}$. Since $H^T Y$ is just a linear tranformation on Y, it follows that $H^T Y$ has mean zero, and variance covariance matrix $H^T \Sigma H$. Hence using the definition of REML,

$$L_{\text{REML}} = e^{-Y^T H (H^T \Sigma H)^{-1} H^T Y/2} \Big/ \left|H^T \Sigma H\right|^{1/2}. \tag{2.6}$$

We will use the following identities:

$$H(H^T \Sigma H)^{-1} H^T = \Sigma^{-1} - \Sigma^{-1} X (X^T \Sigma^{-1} X)^{-1} X^T \Sigma^{-1} \tag{2.7}$$

(see Searle, Casella and McCulloch, 1992, Appendix M4.f), and

$$\left|H^T \Sigma H\right|^{1/2} \propto |\Sigma|^{1/2} \left|X^T \Sigma X\right|^{1/2}. \tag{2.8}$$

This last identity is proved at the end of this section.

From (2.7) it follows that

$$\begin{aligned} Y^T H (H^T \Sigma H)^{-1} H^T Y &= Y^T \Sigma^{-1} Y - Y^T \Sigma^{-1} X (X^T \Sigma^{-1} X)^{-1} X^T \Sigma^{-1} Y \\ &= (Y - X\widehat{\beta})^T \Sigma^{-1} (Y - X\widehat{\beta}) \\ &= \Sigma_i (Y_i - X_i \beta)^T \Sigma^{-1} (Y_i - X_i \beta). \end{aligned}$$

where

$$\widehat{\beta}(\Sigma^{-1}) = (X^T \Sigma^{-1} X)^{-1} X^T \Sigma^{-1} Y.$$

Thus from (2.6, 2.7, 2.8) and using the block diagonal structure of Σ we have

$$L_{\text{REML}} = e^{-\text{tr}(S\Sigma^{-1})/2} \Big/ |\Sigma|^{1/2} \left|\Sigma_i X_i \Sigma^{-1} X_i\right|^{1/2}$$

for $S = \Sigma_i (Y_i - X_i \widehat{\beta})(Y_i - X_i \widehat{\beta})^T$. It follows immediately that the previous two characterizations of the REML likelihood are also true:

1. The REML likelihood is the profile likelihood, $L(\beta = \widehat{\beta}_{\text{ML}}, \Sigma)$, multiplied by $|\text{var}\widehat{\beta}(\Sigma)|^{1/2}$.

2. The REML likelihood is the likelihood of the full sample Y, conditional on $\widehat{\beta}(\Sigma)$.

These characterizations point to a drawback of REML inference. While the profile likelihood is invariant under reparameterizations, the term $\ln |X^T \not\Sigma^{-1} X|$, being a function of the design matrix, is not invariant under reparameterizations of the parameter β. This has the consequence that inference about Σ using REML will change with different, but equivalent, formulations of the linear model. This drawback, however, is asymptotically negligible provided the dimension of β is not too large. For further discussion see Bardnorff-Nielsen and Cox (1994, Section 4.4).

The use of the REML likelihood can also be based on marginal sufficiency arguments given in Sprott (1975). That is, $H^T Y$ is marginally sufficient because:

1. the distribution of $H^T Y$ depends only on Σ,
2. the distribution of $\widehat{\beta}$ contains no information about Σ in the absence of knowledge about β, and
3. together $H^T Y$ and $\widehat{\beta}$ are a full rank transformation of the data Y.

See also Kalbfleish and Sprott (1970, 1973, 1976).

It is straightforward to show that the REML score equations for Σ are

$$N\Sigma - \sum_{i=1}^{N} X_i \left(\sum_{i=1}^{N} X_i^T \Sigma^{-1} X_i \right)^{-1} X_i^T - S(\Sigma) = 0.$$

so that $\widehat{\Sigma}_{\mathrm{REML}}$ satisfies

$$\widehat{\Sigma}_{\mathrm{REML}} = \frac{1}{N} \sum_{i=1}^{N} \left\{ (Y_i - X_i \, \widehat{\beta}_{\mathrm{REML}})(Y_i - X_i \, \widehat{\beta}_{\mathrm{REML}})^T + X_i V_{\widehat{\beta}} \, X_i^T \right\}$$

where

$$\widehat{\beta}_{\mathrm{REML}} \equiv \widehat{\beta}\left(\widehat{\Sigma}_{\mathrm{REML}}\right) \quad \text{and} \quad V_{\widehat{\beta}} = \left(\sum_{i=1}^{N} X_i^T \, \widehat{\Sigma}_{\mathrm{REML}} \, X_i \right)^{-1}.$$

The iterative computational algorithm given for ML in Section 2.1 can be readily adapted in this setting. Specifically, at stage k, $\widehat{\beta}_k = \widehat{\beta}(\widehat{\Sigma}_k^{-1})$

where $\widehat{\Sigma}_0$ is an arbitrary positive definite matrix but now

$$\widehat{\Sigma}_{k+1} = \sum_{i=1}^{N} \Bigg[(Y_i - X_i \widehat{\beta}_k)(Y_i - X_i \widehat{\beta}_k)^T + X_i \left(\sum_{i=1}^{N} X_i^T \widehat{\Sigma}_k^{-1} X_i \right)^{-1} X_i^T \Bigg] \Big/ N.$$

As with ML estimation, the sequence $(\widehat{\beta}_k, \widehat{\Sigma}_k)$ need not converge, but when it does, the limit of $\widehat{\Sigma}_k$, $k = 1, 2, \ldots$, is a solution of the REML score equations.

The question naturally arises as to which of the ML or REML estimators is to be preferred. As indicated in the introduction, the REML analysis is motivated by the desire to adjust the ML estimator of Σ for lost degrees of freedom. More precisely, the hope is to find an estimator that, like the MLE of Σ, is asymptotically efficient, but in contrast to the MLE, is an unbiased estimator of Σ. The REML estimator of Σ is asymptotically efficient (see, e.g., Cressie and Lahiri, 1993). However, quite generally, the REML estimator need not be unbiased. In Section 3.6 we discuss special cases where the REML estimate is unbiased (and ML is not). In addition, we present examples where the REML estimator has a closed analytical expression.

Proof of identity in (2.8).

To show that

$$|\Sigma| \left| X^T \Sigma^{-1} X \right| \propto \left| H^T \Sigma H \right|$$

where Σ, X, H and Y were defined in Section 2.3, we will define a full rank transformation on Y as TY, where

$$T = \begin{pmatrix} H^T \\ G \end{pmatrix} \quad \text{so that } TY = \begin{pmatrix} H^T Y \\ GY \end{pmatrix}$$

and we choose $G = (X^T \Sigma^{-1} X)^{-1} X^T \Sigma^{-1}$, so that $GY = \widehat{\beta}(\Sigma) \equiv \widehat{\beta}$.

It is easily checked that $H^T \Sigma G = 0$, so $T\ \Sigma T^T$ is block diagonal, and hence

$$\left| T \Sigma T^T \right| = \left| H^T \Sigma^{-1} H \right| \left| (X^T \Sigma^{-1} X)^{-1} \right|.$$

However we also have that $\left| T\ \Sigma T^T \right| = |\Sigma| \left| T T^T \right|$, and using formulas for determinants of a partitioned matrix (Harville, 1997, pp. 188–189), we have

$$\det \begin{pmatrix} H^T H & H^T G^T \\ GH & GG^T \end{pmatrix} = \left| H^T H \right| \left| GG^T - GH(H^T H)^{-1} H^T G^T \right|.$$

Now recall that $H^T H = I$ and $HH^T = I - X(X^T X)^{-1} X^T$, hence

$$\begin{aligned} \left|TT^T\right| &= \left|G(I - HH^T)G^T\right| = \left|GX(X^T X)^{-1} X^T G^T\right| \\ &= \left|X^T X\right|^{-1}, \end{aligned}$$

whence it follows that

$$\left|H^T \Sigma H\right| \left|X^T \Sigma^{-1} X\right|^{-1} = |\Sigma| \left|X^T X\right|^{-1},$$

and $\left|H^T \Sigma H\right| \propto |\Sigma| \left|X^T \Sigma^{-1} X\right|$.

2.4 REML Estimation: A Bayes Approach

The REML estimator of Σ can also be given a Bayesian interpretation (Searle, 1987, Section 9.2) as follows. Suppose we assign both β and Σ uniform (improper) prior distributions over their respective sample spaces. The joint posterior distribution of (β, Σ) is proportional to the likelihood for (β, Σ), where the constant of proportionality depends on the data but not on (β, Σ). Letting $p_s(\beta, \Sigma)$ denote the joint posterior distribution of (β, Σ), we thus have $p_s(\beta, \Sigma) \propto \mathcal{L}(\beta, \Sigma)$. To eliminate the nuisance parameter β, we base inference about Σ on the marginal posterior distribution of Σ, $p_s(\Sigma)$, obtained by integrating $p_s(\beta, \Sigma)$ over β:

$$p_s(\Sigma) \propto \int \mathcal{L}(\beta, \Sigma)\, d\beta. \tag{2.9}$$

To show $p_s(\Sigma)$ is proportional to the REML likelihood, first write

$$\begin{aligned} \mathcal{L}(\beta, \Sigma) \propto |\Sigma|^{-N/2} \exp\Bigg\{ &-\frac{1}{2} \sum_{i=1}^{N} (Y_i - X_i \widehat{\beta})^T \Sigma^{-1} (Y_i - X_i \widehat{\beta}) \\ &+ (\widehat{\beta} - \beta)^T \left(\sum_{i=1}^{N} X_i^T \Sigma^{-1} X_i \right) (\widehat{\beta} - \beta) \Bigg\} \end{aligned}$$

where $\widehat{\beta} = \widehat{\beta}(\Sigma^{-1})$. Then use the identity

$$\int \exp\left\{ -\frac{1}{2} t^T \Omega^{-1} t \right\} db = (2\pi)^{n/2} |\Omega|^{1/2},$$

where Ω is any $n \times n$ positive definite matrix, to obtain that

$$p_s(\Sigma) \propto |\Sigma|^{-N/2} \exp\left\{ \mathrm{tr} \Sigma^{-1} S(\Sigma)/2 \right\} \left| \sum_{i=1}^{N} X_i^T \Sigma^{-1} X_i \right|^{-1/2},$$

so that right hand side is precisely the REML likelihood. Thus, inferences based on the REML likelihood and inferences based on $p_s(\Sigma)$ are identical. In particular, the REML estimate of Σ is the posterior mode. Note that we can alternatively view the REML likelihood as an integrated likelihood where we have integrated out the nuisance parameter β.

2.5 Patterned Σ

When Σ is a patterned matrix, that is $\Sigma = \Sigma(\theta)$ is a function of a vector of parameters θ of dimension less than $n(n+1)/2$, both the ML and REML score equations for θ can be obtained using the chain rule to take derivatives with respect to θ. See, for example, Jennrick and Schluchter (1986). There are several computing packages which provide subroutines for computing ML and REML estimators of θ under a variety of model choices for $\Sigma(\theta)$.

2.6 Closed Form Solutions for $\widehat{\beta}_{\mathbf{ML}}$, $\widehat{\Sigma}_{\mathbf{ML}}$ and $\widehat{\Sigma}_{\mathbf{REML}}$.

For certain design matrices, one can obtain closed form solutions for $\widehat{\beta}_{\text{ML}}$ and $\widehat{\Sigma}_{\text{ML}}$ or $\widehat{\Sigma}_{\text{REML}}$ in the setting where Σ is unstructured and all subjects are measured at all n occasions. Studying these cases is instructive since the resulting estimators show when $\widehat{\beta}_{\text{ML}}$ does not depend on Σ and enables us to quantify the bias of $\widehat{\Sigma}_{\text{ML}}$ and $\widehat{\Sigma}_{\text{REML}}$.

Case 1. The simplest case in which closed form solutions exist corresponds to the classical MANOVA or multivariate regression setting discussed in Section 1.2.3. Furthermore, in this setting the REML estimator of Σ is unbiased. Under the MANOVA model of Section 1.2.3, each $X_i = a_i^T \otimes I_n$, where a_i is a $k \times 1$ vector of subject specific covariates, I_n is an $n \times n$ identity matrix, and $\otimes$ is Kroneker matrix product (Harville, 1999). Effectively each of the n variables has a separate regression model with the same k predictors for each outcome. Let A denote the $N \times k$ matrix with rows given by the a_i^T, i.e., A is the usual design matrix for a single response variable.

For this design, it straightforward to show (see, e.g., Johnson and

Wichern, 1992) that

$$\widehat{\beta}_{\mathrm{ML}} = \widehat{\beta}_{\mathrm{OLS}} = \left(\sum_{i=1}^{N} X_i^T X_i\right)^{-1} \sum_{i=1}^{N} X_i^T Y_i$$

and hence

$$\mathrm{var}\left(\widehat{\beta}_{\mathrm{ML}}\right) = \left(\sum_{i=1}^{N} X_i^T X_i\right)^{-1} \sum_{i=1}^{N} X_i^T \Sigma\ X_i \left(\sum_{i=1}^{N} X_i^T X_i\right)^{-1}$$

Since $E(Y_i) = X_i\beta$, it follows that $\widehat{\beta}_{\mathrm{ML}}$ is unbiased. To compute the bias of $\widehat{\Sigma}_{\mathrm{ML}}$, we will use the following matrix identities:

(i) $\left(a_i^T \otimes I_n\right)^T = a_i \otimes I_n$,

(ii) for conformable matrices A, B, C, D, $(A \otimes B)(C \otimes D) = AC \otimes BD$,

and

(iii) for nonsingular matrices A and B $(A \otimes B)^{-1} = A^{-1} \otimes B^{-1}$.

Also from Johnson and Wichern (1990) we have

$$\widehat{\Sigma}_{\mathrm{ML}} = \frac{1}{N}\sum_{i-1}^{N}(Y_i - X_i\widehat{\beta}_{\mathrm{ML}})(Y_i - X_i\widehat{\beta}_{\mathrm{ML}})^T,$$

thus,

$$\begin{aligned}\widehat{\Sigma}_{\mathrm{ML}} = & \Bigg[\sum_{i=1}^{N}(Y_i - X_i\beta)(Y_i - X_i\beta)^T + \sum_{i=1}^{N} X_i(\widehat{\beta}_{\mathrm{ML}} - \beta)(\widehat{\beta}_{\mathrm{ML}} - \beta)X_i^T \\ & -\sum_{i=1}^{N}(Y_i - X_i\beta)(\widehat{\beta}_{\mathrm{ML}} - \beta)^T X_i^T \\ & -\sum_{i=1}^{N} X_i(\widehat{\beta}_{\mathrm{ML}} - \beta)(Y_i - X_i\beta)^T\Bigg] \Big/ N.\end{aligned}$$

Taking expectations and setting $X_i = a_i^T \otimes I$, we find

$$E(\widehat{\Sigma}_{\mathrm{ML}}) = \Sigma - \frac{1}{N}\sum_{i=1}^{N} X_i\ \mathrm{var}\left(\widehat{\beta}_{\mathrm{ML}}\right) X_i^T. \tag{2.10}$$

Using the matrix identities given above, one can show that

$$\operatorname{var} \widehat{\beta}_{\mathrm{ML}} = (A^T A)^{-1} \otimes \Sigma$$

and thus

$$E\left(\widehat{\Sigma}_{\mathrm{ML}}\right) = \Sigma - \frac{1}{N}\left(\sum_{i=1}^{N} a_i^T (A^T A)^{-1} a_i\right) \otimes \Sigma.$$

But

$$\sum_{i=1}^{N} a_i^T (A^T A)^{-1} a_i = \operatorname{tr}\left\{\left(\sum_{i=1}^{N} a_i a_i^T\right)(A^T A)^{-1}\right\} = \operatorname{tr}(I) = k,$$

so that

$$E\left(\widehat{\Sigma}_{\mathrm{ML}}\right) = \Sigma(1 - k/N).$$

Thus $\widehat{\Sigma}_{\mathrm{ML}}$ is biased, with bias going to zero as $k/N \to$ zero.

Since $\widehat{\beta}_{\mathrm{ML}}$ does not depend upon $\widehat{\Sigma}_{\mathrm{ML}}$, $\widehat{\beta}_{\mathrm{REML}} = \widehat{\beta}_{\mathrm{ML}} = \widehat{\beta}_{\mathrm{OLS}}$ and

$$\widehat{\Sigma}_{\mathrm{REML}} = \widehat{\Sigma}_{\mathrm{ML}} + \frac{1}{N}\sum_{i=1}^{N} X_i \operatorname{var}\left(\widehat{\beta}_{\mathrm{ML}}\right) X_i^T.$$

and hence from (2.10) it follows that $E(\widehat{\Sigma}_{\mathrm{REML}}) = \Sigma$.

Case 2. The balanced and complete growth curve model discussed in Example 1.2.4 provides another example where the ML and REML estimators have closed form analytical expressions. In this case,

$$X_i = a_i^T \otimes Z$$

where a_i is as before and Z is some suitably defined $n \times q$ design matrix specifying a "design on time" as discussed in Example 1.2.4. Recall that in this setting we may write:

$$E(Y_i) = X_i \beta = Z \Delta a_i$$

with Δ defined as in Section 1.2.4. Grizzle and Allen (1969) (see also Khatri, 1966) have obtained a closed form solution for Δ in this case as

$$\widehat{\Delta}_{\mathrm{ML}} = (Z^T S^{-1} Z)^{-1} Z^T S^{-1} Y A (A^T A)^{-1} \tag{2.11}$$

where Y is now an $n \times N$ data matrix whose columns are Y_i, and

$$S = Y\left[I - A(A^T A)^{-1} A^T\right] Y^T,$$

where I denotes the identity matrix with dimension n. Notice that we may also write

$$\widehat{\beta}_{\mathrm{ML}} = \mathrm{vec}\ \widehat{\Delta}_{\mathrm{ML}} = \left[(A^T A)^{-1} A^T \otimes (Z^T S^{-1} Z)^{-1} Z^T S^{-1}\right] \ \mathrm{vec}\,(Y)\,.$$

The Grizzle and Allen (1969) derivation is instructive and leads to a proof that $\widehat{\beta}_{\mathrm{ML}}$ is unbiased even though $\widehat{\beta}_{\mathrm{ML}}$ depends upon S. The approach is to make a linear transformation from Y_i to B_i of the form

$$\begin{pmatrix} B_{1i} \\ B_{2i} \end{pmatrix} = \begin{pmatrix} Z_1^T \\ \\ Z_2^T \end{pmatrix} Y_i$$

where Z_1 and Z_2 are full rank matrices of dimension $n \times r$ and $n \times n$, with $n = r + q$, selected so that

$$Z_1^T Z = I$$

and

$$Z_2^T Z = 0.$$

Notice that a natural choice for Z_1 is $Z(Z^T Z)^{-1}$, and for Z_2 is a matrix, $\{I - Z(Z^T Z)^{-1} Z^T\}$, with columns where $v_j, j = 1, \ldots, r$, are $n \times 1$ vectors linearly independent with the columns of Z. Since $E(Z_2 Y_i) = 0$, the marginal distribution of $Z_2 Y_i$ does not depend upon β, hence we can write

$$\begin{aligned} \Pi\ f(Y_i|\beta,\Sigma) &\propto \ \Pi\ f(B_i|\beta,\Sigma) \\ &= \Pi\ f(B_{1i}|\beta,\Sigma,B_{2i}) f(B_{2i}|\Sigma) \end{aligned}$$

and the MLE of β is obtained by maximizing the first component only. Because B_{1i} and B_{2i} are jointly multivariate normally distributed, the conditional mean of B_{1i} given B_{2i} is linear in B_{2i}. Thus, letting

$$\begin{aligned} \theta &= \mathrm{cov}(B_{1i}, B_{2i})(\mathrm{var}\ B_{2i})^{-1} \\ &= Z_1^T \Sigma Z_2 (Z_2^T \Sigma Z_2)^{-1}. \end{aligned}$$

we have

$$\begin{aligned} E(B_{1i}|B_{2i};\Sigma,\beta) &= \Delta a_i + \theta B_{2i} \\ &= [\Delta, \theta] \begin{pmatrix} a_i \\ B_{2i} \end{pmatrix} \\ &= \tau\ W_i, \end{aligned}$$

where Δ is defined as in Section 1.2.4, τ is a $q \times (k+r)$ matrix of parameters, and W_i is a $(k+r) \times 1$ vector of covariates. But this is identical to our model in Case 1 where we write

$$E(B_{1i}) = (W_i^T \otimes I)\Omega$$

and Ω =vec(τ). Hence the ML estimate of τ is OLS, and can be written as

$$\widehat{\tau} = B_1 W (W^T W)^{-1}$$

where B_1 is the $q \times N$ data matrix whose columns are B_{1i} and W is the $N \times (k+r)$ design matrix whose rows are W_i^T. By partitioning $\widehat{\tau}$ and using standard matrix identities, the result for $\widehat{\Delta}$ in (2.11) is obtained. It also follows that $E(\widehat{\Delta}|B_2; \Delta, \Sigma) = \Delta$ where B_2 is the vector with components B_{2i}, $i = 1, \ldots, N$, and therefore $\widehat{\Delta}$ is also unconditionally unbiased for Δ.

Case 3. These results concerning closed form solutions have been extended by Szatrowski (1980), Szatrowski and Miller (1980) and Lange and Laird (1989) for the setting where Σ takes a random effects structure, i.e.,

$$\Sigma = Z_c D Z_c + \sigma^2 I,$$

where Z_c is a subset of the columns of Z of dimension $n \times c$, D is a positive definite matrix and I is the $n \times n$ identity matrix. Here it can be shown that $\widehat{\beta}_{\text{ML}} = \widehat{\beta}_{\text{REML}} = \widehat{\beta}_{\text{OLS}}$, and simple closed form expressions can be derived for the ML and REML estimates of D and σ^2. Lange and Laird (1989) showed also that the REML estimates of Σ are unbiased and the ML estimates have bias which goes to zero as $k/N \to 0$, for fixed q.

Chapter 3

Estimation in the LMCD Assuming Normally Distributed Errors with Unbalanced Designs/Missing Data

In Chapter 2 we considered only designs with equal numbers of observations, n, on each subject. It often happens that $n_i \neq n$ and hence $\text{var}(Y_i) = \sum_{i\,(n_i \times n_i)}$; this can arise in different settings. We can have studies with unequal n_i by design: e.g., clustering, sampling households or litters, or family studies. In this case the Y_i are complete responses on the sampling unit, and the LMCD assumes

$$\underset{n_i \times 1}{E(Y_i)} = \underset{n_i \times p}{X_i} \underset{p \times 1}{\beta} \tag{3.1}$$

and

$$\text{var}(Y_i) = \underset{n_i \times n_i}{\Sigma_i}, \tag{3.2}$$

for appropriate choices of X_i, β and Σ_i. Here Σ_i depends on i through its dimension, and possibly also X_i, but we will assume a common parameter set for the Σ_i, so that we may write $\Sigma_i(\theta)$, where θ contains all of the variance covariance parameters. For example, if each observation has the same variance σ^2 and any two pairs of Y_{ij}, Y_{ik} have the same covariance, then each Σ_i has the compound symmetry form with different dimension n_i.

Alternately, it may happen that the design calls for n measures per subject, with

$$\underset{n\times 1}{E\,(Y_i)} = \underset{n\times p}{X_i}\;\underset{p\times 1}{\beta}$$

and

$$\text{var}\,(Y_i) = \underset{n\times n}{\Sigma_i}\,,$$

but some observations are missing, so that the vector on n_i observations that are the observed data can be expressed as

$$\underset{n_i\times n}{Y_i^{\text{OBS}}} = \underset{n_i\times n}{I_i}\;\underset{n\times 1}{Y_i}\,.$$

where I_i is obtained from an $n \times n$ identity matrix by removing the rows corresponding to the missing observations. Here again the dimension of each Σ_i will depend upon i; as before we assume a common parameter vector θ which consists of the unique elements of Σ. In general,

$$E\left(Y_i^{\text{OBS}}\right) = I_i\,E(Y_i) = I_i\,X_i\,\beta = X_i^{\text{OBS}}\,\beta \tag{3.3}$$

and

$$\text{var}\left(Y_i^{\text{OBS}}\right) = I_i\,\Sigma_i\,I_i^T = \Sigma_i^{\text{OBS}} \tag{3.4}$$

hold only if the MDM is MCAR. It is important to note that the β and θ are the same for the unequal n_i and equal n case. From a technical point of view, both of the cases (unbalanced by design or because of MCAR data) can be handled in the same way (although the structures for Σ_i will generally be different), but the validity of the estimators will depend upon the design or the MDM. Henceforth we drop the superscript OBS and all Y_i are $n_i \times 1$ and X_i are $n_i \times p$, unless otherwise noted.

3.1 ML and REML Estimation for the Unequal n_i Case

Here we will consider ML and REML estimation based on the multivariate normal distribution for unbalanced data. As discussed above, this can be appropriate for the clustered data setting, or it can arise in the case of missing data. In this chapter we deal with likelihood inference for MAR and/or MCAR MDM's, hence we can assume that each observed Y_i is $N(X_i\beta, \Sigma_i)$. As in the complete data case, we will mainly focus on unstructured Σ in this chapter, i.e., the missing data case where the covariance matrix for each complete data vector is Σ. Chapter 5 takes up a random effects structure which allows Σ to depend on X_i.

Letting θ denote the parameters of Σ_i, we then have that the likelihood based on the observed data is

$$\mathcal{L}(\beta,\theta) = \prod_{i=1}^{N} |\Sigma_i|^{-1/2} \exp\left[-\frac{1}{2}\sum_{i=1}^{N}(Y_i - X_i\beta)^T\, \Sigma_i^{-1}\, (Y_i - X_i\beta)\right].$$

Again, it is easily seen that for any fixed θ (and Σ_i), the ML estimate of β is given by

$$\widehat{\beta}(\theta) = \left(\sum_{i=1}^{N} X_i^T \Sigma_i^{-1}(\theta) X_i\right)^{-1} \sum_{i=1}^{N} X_i^T \Sigma_i^{-1}(\theta) Y_i. \tag{3.5}$$

Thus, as in the complete data case, the ML estimate of β has a simple form as a function of θ.

Substituting $\widehat{\beta}(\theta)$ for β in the likelihood gives the profile likelihood

$$\mathcal{L}(\widehat{\beta},\theta) = \prod_{i=1}^{N} |\Sigma_i|^{-1/2} \exp\left[-\frac{1}{2}\sum_{i=1}^{N}(Y_i - X_i\widehat{\beta}(\theta))^T \Sigma_i^{-1}(Y_i - X_i\widehat{\beta}(\theta))\right].$$

Even when $\Sigma_i = \underset{n_i\times n}{I_i}\ \underset{n\times n}{\Sigma}\ \underset{n\times n_i}{I_i^T}$ for Σ unstructured, the derivation of the score equation is complex and there are not simple closed form solutions to the score equations for Σ. A set of simple iterative computing equations can be derived using the EM algorithm and we can also use the "EM-method" to derive the score equations for Σ. This will be described in Section 3.3; for now, we give the following expression for $\widehat{\Sigma}_{\text{ML}}$ in the case of missing data and $\Sigma_i = I_i\, \Sigma\, I_i^T$:

$$\widehat{\Sigma}_{\text{ML}} = \sum_{i=1}^{N} \widehat{Q}_i / N \tag{3.6}$$

where

$$Q_i = \Sigma - \Sigma\, I_i^T\, \Sigma_i^{-1}\, I_i\, \Sigma + \sum_{i=1}^{N} \Sigma I_i^T\, \Sigma_i^{-1}\, (Y_i - X_i\beta)\, (Y_i - X_i\beta)^T\, \Sigma_i^{-1}\, I_i^T\, \Sigma,$$

$$\beta = \left(\sum_{i=1}^{N} X_i^T\, \Sigma_i^{-1}\, X_i\right)^{-1} \sum_{i=1}^{N} X_i^T\, \Sigma_i^{-1}\, Y_i,$$

$$\Sigma_i = I_i\, \Sigma\, I_i^T,$$

and $\widehat{Q}_i$ is Q_i with $\widehat{\Sigma}$, $\widehat{\Sigma}_i$ and $\widehat{\beta}(\widehat{\Sigma}_i^{-1})$ substituted for Σ, Σ_i and β.

The derivation of the REML estimator of θ proceeds as in the complete data case, by constructing the profile likelihood, with an additional term in the denominator for $\widehat{\text{var}(\beta)}$:

$$\mathcal{L}_k(\theta) = \prod_{i=1}^{N} |\Sigma_i|^{-1/2} \left| \sum_{i=1}^{N} X_i^T \Sigma_i^{-1} X_i \right|^{-1/2} \times \exp\left[-\frac{1}{2} \sum_{i=1}^{N} (Y_i - X_i\widehat{\beta})^T \Sigma_i^{-1} (Y_i - X_i\widehat{\beta}) \right],$$

As before, with unstructured Σ we can use the EM approach explained in (3.3) to derive a solution to the likelihood equations

$$\widehat{\Sigma}_{\text{REML}} = \sum_{i=1}^{N} (\widehat{Q}_i + \widehat{T}_i)/N \tag{3.7}$$

where Q_i is as before,

$$T_i = \Sigma\, I_i^T\, \Sigma_i^{-1}\, X_i \left(\sum_{i=1}^{N} X_i^T \Sigma_i^{-1} X_i \right)^{-1} X_i^T\, \Sigma_i^{-1}\, I_i\, \Sigma,$$

and $\widehat{Q}_i$ and $\widehat{T}_i$ are Q_i and T_i evaluated at $\widehat{\Sigma}_{\text{REML}}$. For both the ML and REML estimators, simple iterative computing algorithms will be derived in Section 3.6.

Notice that setting $I_i = I$ and $\Sigma_i = \Sigma$ for all i (no missing data) gives the corresponding likelihood equations for the complete data case with unstructured Σ. From (3.6)–(3.7) it is clear that $\widehat{\Sigma}_{\text{ML}}$ and $\widehat{\Sigma}_{\text{REML}}$, will be consistent for Σ even if normality does not hold, provided the data are MCAR.

Under an MAR assumption on MDM, equations (3.1)–(3.2) do not hold, i.e.,

$$\underset{n_i \times 1}{E\,(Y_i)} \neq \underset{n_i \times p}{X_i}\ \underset{p \times 1}{\beta}$$

and

$$\text{var}\,(Y_i) \neq \Sigma_i = I_i \Sigma I_i^T,$$

where β and Σ determine the moments of the complete data vectors. However, with data MAR, we saw in Section 1.4 that the appropriate likelihood for (β, θ) is proportional to the likelihood based on the ordinary marginal of the observed data, i.e., the likelihood obtained by assuming $Y_i \sim N(X_i\beta, \Sigma_i)$. Thus the likelihood estimators for (β, θ) do not differ for the MAR and MCAR case even though their properties differ under the different missingness assumptions. So, for $\widehat{\beta}_{\mathrm{ML}}$ defined as

$$\widehat{\beta}_{\mathrm{ML}} = \left(\sum_{i=1}^{N} X_i^T \widehat{\Sigma}_i^{-1} X_i \right)^{-1} \sum_{i=1}^{N} X_i^T \widehat{\Sigma}_i^{-1} Y_i,$$

where $\widehat{\Sigma}_i$ is also based on $\widehat{\theta}_{\mathrm{ML}}$, we have that $\widehat{\beta}_{\mathrm{ML}}$ is a consistent estimate of β and $\widehat{\Sigma}_{\mathrm{ML}}$ and $\widehat{\Sigma}_{\mathrm{REML}}$ are both consistent for Σ in the case of missing data if (a), (b) and (c) or (d) hold:

1. (a) $E(\underset{n\times 1}{Y_i}) = \underset{n\times p}{X_i}\ \underset{p\times 1}{\beta}$ holds for the complete case.

   (b) $\underset{n\times n}{\Sigma} = \mathrm{var}(\underset{n\times 1}{Y_i})$ holds for the complete case.

   (c) Error terms are normal and the missingness is MAR.

   or

   (d) The missingness is MCAR.

   Note also that:

2. Under (a)–(c), the observed Fisher Information Matrix, but not the expected information, gives a valid estimate of var $(\widehat{\beta}_{\mathrm{ML}})$. So Avar $(\widehat{\beta}_{\mathrm{ML}}) \neq (\sum_{i=1}^{N} X_i^T \widehat{\Sigma}_i^{-1} X_i)^{-1}$ except with MCAR. In practice, however, there is often little difference between the estimated observed and expected information.

3. Under MCAR, $\widehat{\beta}_{\mathrm{ML}}$ is unbiased for β (Karkar and Harville, 1988).

3.2 A General Formulation for Incomplete Data

As mentioned in Section 3.1, using straightforward matrix differentiation techniques to obtain the derivatives of the profile likelihood for θ is tedious, and does not lead to any particularly tractable set of equations. An alternative method for deriving likelihood equations is to rely on a general theory for maximum likelihood in the presence of incomplete data, which includes missingness as well as imbalance by design. The general theory is also the motivation for the EM algorithm.

The EM algorithm is a very general algorithm for computing ML and REML estimates with incomplete data (Dempster, Laird and Rubin, 1977). Indeed, by appropriately defining the complete data, the algorithm can also be applied to solve the ML and REML score equations in a broad class of LMCD models with normal errors (like the random effects model discussed in the Chapter 5), *even in the absence of missing data.* In this section, we introduce a general notation for describing incomplete data.

Let $\mathcal{Y}$ be an observed data vector which can be represented as an incomplete version of some complete data vector $\mathcal{Z}$ with density $f_{\mathcal{Z}}(z; \Phi)$: $\mathcal{Y}$ may be simply missing components of $\mathcal{Z}$, or it may be a convolution of components of $\mathcal{Z}$. By definition, the density of $\mathcal{Y}$, $f_{\mathcal{Y}}(y; \Phi)$, satisfies

$$f_{\mathcal{Y}}(y; \Phi) = \int f_{\mathcal{Z}}(z; \Phi)\, dz_y, \tag{3.8}$$

where dz_y denotes integration over the missing (or incomplete) data.

We will illustrate the algorithm with four examples. All of the examples are special cases of the LMCD where $\Phi = (\beta, \Sigma)$. In all of these cases, $\widehat{\beta}_{\mathrm{ML}}$ based on the observed data is given by (3.6) for a given Σ. The difficulty is in deriving score equations for Σ and computing $\widehat{\Sigma}_{\mathrm{ML}}$. It helps to keep this in mind when reviewing the examples, since the incomplete data problem is formulated to make ML estimation of Σ easy. The goal is to compute the maximum likelihood estimator of some or all elements of Φ. The observed data $\mathcal{Y}$ may be a pre-specified subset of $\mathcal{Z}$ or it may be a random subset.

Example 1. Suppose each $Z_i \sim N_n(\mu, \Sigma)$, $i = 1, \ldots, N$, and some subjects are missing some of the Z_{ij}'s. Let Y_i denote the $(n_i \times 1)$ subset of the Z_{ij}'s which are observed. Then $\mathcal{Y}^T = (Y_1^T, \ldots, Y_N^T)$ is a $n_+ \times 1$ vector, where n_+ is the summation of the n_i, and we will let the complete data $\mathcal{Z}$ be the $nN \times 1$ vector defined as $\mathcal{Z}^T = (Z_1^T, \ldots, Z_N^T)$. Note this is a special case of the LMCD when $X_i = I$ and $\mu = \beta$.

Example 2. Let $Z_i \sim N_n(X_i\beta, \Sigma)$ where β and Σ are unknown and Σ is an arbitrary symmetrc positive definite matrix. Take $\mathcal{Y}^T = (Z_1^T, \ldots, Z_N^T)$ and $\mathcal{X}^T = (Z_1^T, e_1^T, \ldots, Z_N^T, e_N^T)$. Here the outcomes Z_{ij} are fully observed. However, we regard the unobserved residuals, $e_i = (Y_i - X_i\beta)$, as additional, unobserved data. Example 2 seems a bit contrived, but provides us with a simple way of deriving the likelihood equations for (β, Σ) as given in Chapter 3, which extends easily when some outcomes Z_{ij} are missing, as indicated in the next example.

Example 3. As in Example 2, let $Z_{i\,(n \times 1)} \sim N_n(X_i\beta, \Sigma)$ but now suppose that some Z_{ij} are missing, so only $Y_{i\,(n_i \times 1)}$ is observed. Then take $\mathcal{Y}^T = (Y_1, \ldots, Y_N)^T$ and $\mathcal{Z}^T = (Y_1^T, e_1^T, \ldots, Y_N^T, e_N^T)$ as in example 2. The dimension of $\mathcal{Z}$ is $(nN \times 1)$ and of $\mathcal{Y}$ is $n_+ \times 1$.

Example 4. Suppose $Y_{ij} = X_{ij}^T\beta + b_i + e_{ij}$, $j = 1, \ldots, n_i$, and $i = 1, \ldots, N$ where b_i and $e_{ij}, j = 1, \ldots, n_i$, are all independent normally distributed with with zero mean, $\text{var}(b_i) = d$ and $\text{var}(e_{ij}) = \sigma^2$. We take $\mathcal{Z}^T = (Y_1^T, b_1, \ldots, Y_N^T, b_N)$ and $\mathcal{Y}^T = (Y_1^T, \ldots, Y_N^T)$. As in Example 2, there are no missing data, and the random effects are considered as missing for convenience. Random effects models will be taken up in detail in Chapter 5.

The first thing to notice is that the definition of the complete data $\mathcal{Z}$ need not be unique. The choice of the observed data $\mathcal{Y}$ is unique because it is actually observed. In general, there may be many choices for $\mathcal{Z}$. We should choose $\mathcal{Z}$ so that computing the MLE of Φ based on $\mathcal{Z}$ is easy. For instance, in example 1 with the complete data $\mathcal{Z} = (Z_1^T, \ldots, Z_N^T)$, the ML estimators of (μ, Σ) have closed form. But in Example 3, where the mean of each Z_i is $X_i\beta$, closed form estimates do not exist except in very special cases. By taking $\mathcal{Z}$ to include the error terms we can get closed form estimates of Σ with complete data. As we will show in Chapter 5, with Example 4, adding only the random effects b_i, means that we can get closed form estimates for (β, d, σ^2) with the complete data.

3.3 Derivatives of the Log-likelihood for the Incomplete Data Model and the EM Algorithm

Derivatives of incomplete data likelihoods can be derived using the following fundamental identity. Suppose that regularity conditions are satisfied so that the order of differentiation and integration can be interchanged in the right hand side of (3.8). Then we have

$$\begin{aligned}
\partial \log f_{\mathcal{Y}}(y;\Phi)/\partial\Phi &= \frac{\int \partial f_{\mathcal{Z}}(z;\Phi)/\partial\Phi \; dz_y}{f_{\mathcal{Y}}(z;\Phi)} \\
&= \frac{\int \partial \log f_{\mathcal{Z}}(z;\Phi)/\partial\Phi \; f_{\mathcal{Z}}(z;\Phi) dz_y}{f_{\mathcal{Y}}(y;\Phi)} \\
&= \int \partial \log f_{\mathcal{Z}}(z;\Phi)/\partial\Phi \; f_{\mathcal{Z}|\mathcal{Y}}(z|y;\Phi) dz_y \\
&= E_{\mathcal{Z}|\mathcal{Y}} \{\partial \log f_{\mathcal{Z}}(z;\Phi)/\partial\Phi | y;\Phi\}
\end{aligned}$$

where the subscript $\mathcal{Z}|\mathcal{Y}$ refers to the conditional distribution of $\mathcal{Z}$ given $\mathcal{Y}$. Thus, we see that the score for Φ based on the observed data $\mathcal{Y} = y$ is equal to the expectation of the complete data score for Φ based on

$\mathcal{Z} = z$, conditional on the observed data. This identity provides us with an easy way of deriving the score equations in incomplete data settings.

In the score equations for Φ,

$$E_{\mathcal{Z}|\mathcal{Y}}\left\{\partial \log f_{\mathcal{Z}}(z;\Phi)/\partial\Phi|y;\Phi\right\} = 0 \tag{3.9}$$

the parameter Φ enters twice; once as an index of the complete data score $\partial \log f_{\mathcal{Z}}(z;\Phi)/\partial\Phi$ and then as an index of the conditional distribution of $\mathcal{Z}$ given $\mathcal{Y}$. This suggests solving (3.9) by an iterative algorithm that updates the values of Φ separately in its two appearances. Specifically, given Φ_k, stage $k+1$ of the algorithm consists of two steps, namely:

E-step Compute the conditional expectation of the complete data score

$$E\left\{\partial \log f_{\mathcal{Z}}(z;\Phi)/\partial\Phi|y;\Phi_k\right\}$$

and

M-step Solve for Φ in the expected complete data score equations

$$E\left\{\partial \log f_{\mathcal{Z}}(z;\Phi)/\partial\Phi|y;\Phi_k\right\} = 0$$

and call its solution Φ_{k+1}.

The E-step (E for expectation) essentially imputes the complete data score using its expectation given the observed data. The M-step (M for maximization) solves the complete data likelihood equations using the expected score. One instance in which the algorithm takes a particularly simple form is when the distribution of the complete data has an exponential family distribution, namely

$$f_{\mathcal{Z}}(z;\Phi) = \frac{e^{t(z)^T\theta(\Phi) - b(z)}}{a\left\{\theta\left(\Phi\right)\right\}},$$

where θ is the vector of natural parameters and $t(x)$ is the corresponding vector of sufficient statistics. Then the complete data score function for Φ is

$$\partial \log f_{\mathcal{Z}}(z;\Phi)/\partial\Phi = \left\{t\left(z\right) - E\left[t\left(z\right);\Phi\right]\right\}^T \left\{\frac{\partial\theta\left(\Phi\right)}{\partial\Phi}\right\},$$

hence

$$E\left\{\partial \log f_{\mathcal{Z}}(z;\Phi)/\partial\Phi|y;\Phi\right\} = \left\{E_{\mathcal{Z}|\mathcal{Y}}\left[t\left(z\right)|y;\Phi\right] - E\left[t\left(z\right);\Phi\right]\right\}^T \left\{\frac{\partial\theta\left(\Phi\right)}{\partial\Phi}\right\}.$$

Thus, in this case given Φ_k, the $(k+1)$th stage of the EM algorithm is:

E-step Compute $t_{k+1}(y) = E_{\mathcal{Z}|\mathcal{Y}} \{t(z) | y; \Phi_k\}$, and

M-step Solve

$$\{t_{k+1}(y) - E[t(z); \Phi]\} \left\{ \frac{\partial \theta(\Phi)}{\partial \Phi} \right\} = 0,$$

or when the transformation from θ to Φ is invertible,

$$t^{(k+1)}(y) = E[t(z); \Phi]. \tag{3.10}$$

Notice that solving for Φ_{k+1} at the M-step is the same as solving the complete data likelihood equations with the sufficient statistics $t(z)$ replaced by its imputation $t_{k+1}(y)$. Also notice that the algorithm assumes the existence of a "complete data vector" $\mathcal{Z}$ but it does not prescribe how to define $\mathcal{Z}$. Ordinarily, one would want to define $\mathcal{X}$ in such a way that the steps of the algorithm can be easily computed. In particular, if $\mathcal{Z}$ has a distribution in an exponential family, one would want to choose $\mathcal{Z}$ so that $E[t(z); \Phi]$ is itself equal to Φ or an easily invertible function of Φ.

We now return to our examples.

Example 1. Deriving the EM equations for Example 1 is very straightforward, and left as an exercise. The M-step has a simple closed form solution for μ_{k+1}, Σ_{k+1} and $E(t(x)|\mathcal{Y}, \Phi_k)$ is easily obtained using standard multivariate theory. See, for example, Little and Rubin (1987), Section 8.2.

Example 2. Now we have no missing data (each Z_i is $n \times 1$) but since each subject has a possibly different mean, the ML and REML estimates of Σ do not have closed form. In Section 2.3 we calculated the score equations for β and Σ directly by derivation of the log-likelihood function. Here we show how the "EM Method" can be used to obtain a solution to the score equations avoiding the need of matrix derivations. Our implementation of the EM for this example is unorthodox; because the complete data likelihood is singular (y_i is linear in e_i), we always use the score equations for β obtained from the observed data and use the complete data to derive score equations for Σ. With $(e_1^T, \ldots, e_N^T)$ in the complete data where each e_i is iid $e_i \sim N(0, \Sigma)$, we have that $S = \sum_{i=1}^{N} e_i e_i^T$ is the minimal sufficient statistic for Σ, and the complete data score equations for Σ are

$$N\Sigma - S. \tag{3.11}$$

Further, the score equations for β, based on the observed data, are

$$\sum_{i=1}^{N} X_i^T \Sigma^{-1}(Z_i - X_i\beta).$$

Then we have the following:

E-step Given Σ_k and β_k, set

$$S_k = \sum_{i=1}^{N} E\left(e_i e_i^T | X_i, \Sigma_k, \beta_k\right) = \sum_{i=1}^{N} (Z_i - X_i\beta_k)(Z_i - X_i\beta_k)^T,$$

because var $(e_i|Z_i, \Sigma) = 0$ and $E(e_i|Z_i, \Sigma) = (Z_i - X_i\beta)$. Note that the score equation for β does not depend upon the e_i's.

M-step Given S_k, compute Σ_{k+1} as

$$\Sigma_{k+1} = S_k/N = \textstyle\sum_{i=1}^{N}(Z_i - X_i\,\beta_k)(Z_i - X_i\,\beta_k)^T/N,$$

and β_{k+1} as

$$\beta_{k+1} = (\Sigma X_i^T \Sigma_{k+1}^{-1} X_i)^{-1} \Sigma X_i^T \Sigma_{k+1}^{-1} Z_i.$$

This is simply iteratively reweighted generalized least squares (IRLS) for the case where we have complete multivariate data. At convergence, $\Sigma_k = \Sigma_{k+1} = \widehat{\Sigma}_{\text{ML}}$ and we obtain the likelihood equations given in Section 3. Because at each iteration, $\widehat{\beta}_{k+1}$ maximizes the observed data likelihood and not the expected complete data likelihood, Liu and Rubin (1994) refer to this as a generalization of the EM.

Example 3. Suppose that in Example 2, some of the Z_{ij}'s are missing. Since $\mathcal{Z}$ is the same, the M-step for Σ remains the same, but now

$$E\,(\underset{n\times 1}{e_i} \,|\, \underset{n_i\times 1}{Y_i}, \beta, \Sigma) = I_i^T\, \Sigma_i^{-1}\, (Y_i - X_i\beta)$$

and

$$\text{var}\,(e_i|Y_i, \beta, \Sigma) = \Sigma - \Sigma\, I_i^T\, \Sigma_i^{-1}\, I_i\, \Sigma,$$

so that

$$E\,(e_i e_i^T|Y_i, \beta, \Sigma) = \Sigma - \Sigma I_i^T I_i \Sigma + \Sigma\, I_i^T \Sigma_i^{-1}(Y_i - X_i\beta)(Y_i - X_i\beta)^T \Sigma_i^{-1} I_i \Sigma = Q_i.$$

As before, using the observed score equation for β yields

$$\sum_{i=1}^{N} X_i^T \Sigma_i^{-1}(Y_i - X_i\beta). \tag{3.12}$$

Again combining the E- and M-steps, for a given β_k and Σ_k we have

$$\Sigma_{k+1} = \sum_{i=1}^{N} Q_{ik} \Big/ N \tag{3.13}$$

where Q_{ik} is Q_i with (β, Σ) evaluated at (β_k, Σ_k). Solving (3.12) for $\beta_{k+1}, \Sigma_{k+1}$ gives

$$\beta_{k+1} = \left(\sum_{i=1}^{N} X_i^T \Sigma_{ik}^{-1} X_i \right)^{-1} \Sigma X_i^T \Sigma_{ik}^{-1} Y_i, \tag{3.14}$$

where Σ_{ik}^{-1} is $(I_i \Sigma I_i^T)^{-1}$ evaluated at Σ_k. Using (3.13)–(3.14) might be regarded as a variant of multivariate IRLS for missing data. Notice that if we have complete data on all subjects the algorithm reduces to the IRLS algorithm of Example 2. From the definition of Q_i as $E(e_i e_i^T | Y_i, \beta, \Sigma)$ and $S = \sum_{i=1}^{N} e_i e_i^T$, it follows from (3.10) and (3.11) that the score equation for $\widehat{\Sigma}_{\mathrm{ML}}$ is given by (3.6).

The EM approach can also be used to derive the REML likelihood equations and the corresponding iterative algorithms, using the Bayes formulation for EM. Recall that with REML, β and Σ are given flat priors, and the REML likelihood is proportional to the marginal posterior of Σ. Thus using (4.5) where $\Phi = (\beta, \Sigma)$ and $\mathcal{Y}$ is the observed data, the REML likelihood is

$$\mathcal{L}_R(y, \Sigma) \propto \int f_{\mathcal{Y}}(y | \beta, \Sigma) d\beta \propto \int \int f_{\mathcal{Z}}(z | \Sigma, \beta) dz_y d\beta.$$

This implies we can regard β as an additional piece of "missing" data. Hence the sufficient statistic for Σ remains S, but in computing the expectations given $\mathcal{Y}$, we consider the β as missing data. Thus the REML score equations become

$$N\Sigma - E(E(S | \mathcal{Y}, \beta, \Sigma))$$

where inner expection is exactly the same as before, and where the outer expectation is with respect to β given $\mathcal{Y} = y$ and Σ.

To evaluate the REML score equations, we use the fact that with linear models, normal likelihoods, and flat priors on β, the posterior of β given Σ and the data is normal, with mean equal to $\widehat{\beta}(\Sigma)$ and variance equal to the variance of $\widehat{\beta}$: $\mathrm{var}(\beta) = (\sum_{i=1}^{N} X_i^T \Sigma_i^{-1} X_i)^{-1}$. Therefore, for Example 3,

$$E(S | \mathcal{Y}, \beta, \Sigma) = \sum_{i=1}^{N} E(Q_i | Y_i, \Sigma)$$

and

$$E(Q_i|Y_i, \Sigma) = Q_i(\widehat{\beta}) + T_i,$$

where Q_i and T_i are defined in Section 4.2 and $Q_i(\widehat{\beta})$ is Q_i with β evaluated at $\widehat{\beta}$. Hence it follows that the REML score equations are

$$\widehat{\Sigma}_{\text{REML}} = \sum_{i=1}^{N} \left(Q_i(\widehat{\beta}) + T_i \right) \Big/ N.$$

Notice that if there is no missing data ($Y_i = Z_i$), $S_i = (Y_i - X_i\,\beta)(Y_i - X_i\,\beta)^T$ and

$$\widehat{\Sigma}_{\text{REML}} = \sum_{i=1}^{N} \Bigg[(Y_i - X_i\,\widehat{\beta})(Y_i - X_i\,\widehat{\beta})^T + X_i \left(\sum_{i=1}^{N} X_i^T\,\widehat{\Sigma}_{\text{REML}}^{-1} X_i \right)^{-1} X_i^T \Bigg] \Big/ N$$

where $\widehat{\beta} = \widehat{\beta}\,(\widehat{\Sigma}_{\text{REML}}^{-1})$. When some Z_{ij} are missing, S_i is replaced by $\mathrm{E}\,(e_i\,e_i^T|Y_i,\,\beta,\,\Sigma) = Q_i$, and we simply add $T_i = X_i(\sum_{i=1}^{N} X_i^T \widehat{\Sigma}_{\text{REML}} X_i)^{-1} X_i^T$ onto Q_i to obtain the corresponding REML estimating equations given above.

Chapter 4

Semi-Parametric Estimation in the Linear Model for Correlated Data

In this chapter we consider inference for the parameters of the LMCD without the assumption of multivariate normal errors. We will present a general theory for weighted least squares (WLS) estimators of β, which are consistent and asymptotically normal (CAN), using only the assumption that $E(Y_i) = X_i\beta$. The estimators will be asymptotically efficient as well, provided a consistent estimate of $\text{var}(Y_i)$ is available. We discuss the asymptotic distribution of WLS (and also ML) estimators when $\text{var}(Y_i)$ is misspecified. For simplicity, we will first consider the case where any imbalance is by design so that the expectation $E(Y_i) = X_i\beta$ holds for the observed data and each n_i is fixed. We will discuss imbalance due to missingness at the end of this chapter. We also assume that each $\text{var}(Y_i) = \Sigma_i$ is a function of a finite parameter vector θ. Furthermore, to simplify the presentation we restrict attention to the setting in which the covariates are stochastic, although most of the results described in this section are valid under appropriate regularity conditions when the covariates are assumed to be fixed numbers.

4.1 Weighted Least Squares Estimators of β

A weighted least squares estimator $\widehat{\beta}$ is defined as the value of β that minimizes the objective function

$$Q_W(\beta) = \sum_{i=1}^{N} (Y_i - X_i\beta)^T W_i (Y_i - X_i\beta)$$

where $W_i = W(X_i)$, $i = 1, \ldots, N$, is an arbitrary positive definite and symmetric $n_i \times n_i$ matrix chosen by the investigator. If a minimum of $Q_W(\beta)$ exists, it must solve

$$\frac{\partial Q_W(\beta)}{\partial \beta} = -2\sum_{i=1}^{N} X_i^T W_i (Y_i - X_i\beta) = 0.$$

This equation has a unique solution at

$$\widehat{\beta} = \left(\sum_{i=1}^{N} X_i^T W_i X_i\right)^{-1} \sum_{i=1}^{N} X_i^T W_i Y_i. \tag{4.1}$$

Because each W_i is positive definite, $Q_W(\beta)$ has a positive definite Hessian at $\widehat{\beta}$, and $Q_W(\beta)$ achieves its global minimum at $\widehat{\beta}$.

The structure of $\widehat{\beta}$ simplifies for special choices of weight matrices. For example, if $W_i \propto I$ for any constant σ, then

$$Q_I(\beta) = \sum_{i=1}^{N}\sum_{j=1}^{n} (Y_{ij} - X_{ij}\beta)^2$$

and therefore $\widehat{\beta}$ minimizing $Q_I(\beta)$ is the ordinary least squares estimator of β based on the data $(Y_{ij}, X_{ij}), i = 1, \ldots, N,\ j = 1, \ldots, n_i$. If in addition, $X_i = X$ for all i, such as for example in the polynomial growth curve model of Examples 3 and 4 in Section 1.2, then $\widehat{\beta}$ simplifies to

$$\widehat{\beta} = (X^T X)^{-1} X^T \overline{Y}$$

where $\overline{Y} = \Sigma Y_i / N$. Thus, $\widehat{\beta}$ agrees with the ordinary least squares estimates of the parameters in the regression of the sample means at each occasion on the rows of X.

4.2 Properties of the Weighted Least Squares Estimator

The estimator $\widehat{\beta}$ is unbiased for any choice of weight function because its conditional expectation given $X_1, \ldots, X_N$ satisfies

$$E\left(\widehat{\beta}\right) = \left(\sum_{i=1}^{N} X_i^T W_i X_i\right)^{-1} \sum_{i=1}^{N} X_i^T W_i X_i \beta = \beta. \tag{4.2}$$

Furthermore, by the Gauss–Markov Theorem the weighted least squares estimator that uses $W_i = \Sigma_i^{-1}$ has the smallest conditional variance given the covariates (in the positive definite sense) among the conditional variances of all estimators that are solutions of equations of the form (4.1) for arbitrary choices of W_i. That is, letting $\widehat{\beta}\,(W)$ and $\widehat{\beta}\,(\Sigma^{-1})$ denote the solutions of the estimating equation (4.1) with W_i arbitrary and with $W_i = \Sigma_i^{-1}$ respectively, we have that

$$\text{var}\left\{\widehat{\beta}\,(\Sigma^{-1})\right\} \leq \text{var}\left\{\widehat{\beta}\,(W)\right\} \tag{4.3}$$

where for any squared matrices A and B, $A \leq B$ if and only if $B - A$ is semipositive definite. The proof of this is straightforward; it involves using the fact that $\text{var}[\widehat{\beta}(\Sigma^{-1}) - \widehat{\beta}(W)] \geq 0$.

Notice that the conditional variance of $\widehat{\beta}$ given $X_1, \ldots, X_N$ is

$$\text{var}\left(\widehat{\beta}\right) = \left(\sum_{i=1}^{N} X_i^T W_i X_i\right)^{-1} \left(\sum_{i=1}^{N} X_i^T W_i \Sigma_i W_i X_i\right) \left(\sum_{i=1}^{N} X_i^T W_i X_i\right)^{-1};$$

it reduces to

$$\text{var}\left(\widehat{\beta}\right) = \left(\sum_{i=1}^{N} X_i^T X_i\right)^{-1} \left(\sum_{i=1}^{N} X_i^T \Sigma_i X_i\right) \left(\sum_{i=1}^{N} X_i^T X_i\right)^{-1}$$

when $W_i = I$ and it reduces to

$$\text{var}\left(\widehat{\beta}\right) = \left(\sum_{i=1}^{N} X_i^T \Sigma_i^{-1} X_i\right)^{-1}$$

when $W_i = \Sigma_i^{-1}$.

By (4.2), the estimator that uses $W_i = I$, i.e., the ordinary least squares estimator of β, is unbiased. However, with correlated data, its variance no longer agrees with the one obtained under independence of the repeated observations made on each subject and it depends on the Σ_i's. Calculating the variance of $\widehat{\beta}$ incorrectly assuming independence can, of course, result in Wald tests of β that do not preserve their nominal level. The Gauss Markov Theorem implies that the choice $W_i = I$ may not be optimal. The optimal choice depends on the unknown covariance matrices Σ_i. This suggests that for optimality purposes one could replace the unknown optimal weight Σ_i by an estimator of it and warrants the study of weighted least squares estimators calculated using weight functions that are computed from the data.

4.3 Weighted Least Squares with Data-Dependent Weight Functions

Suppose that the dependence of the weights W_i on X_i can be written as

$$W_i = W(\theta; X_i)$$

for some fixed $q \times 1$ parameter vector θ. In Section 4.2 we have assumed that W_i was a known function of X_i so that θ was fixed and known. In this section we investigate the properties of weighted least squares estimators that use data-dependent weight functions where θ is replaced by a value $\widehat{\theta}$ computed from the sample. Notice that this formulation includes, but is not restricted to, estimators that use as weight W_i an estimator $\widehat{\Sigma}_i^{-1}$ of Σ_i^{-1}. This particular application is obtained by taking $W(\theta; X_i)$ a constant function of X_i and taking θ to be equal to the unknown parameters indexing the model for the covariance of Y_i.

Henceforth, let $\widehat{\beta}$ be the solution to the equation

$$\sum_{i=1}^{N} X_i^T \widehat{W}_i (Y_i - X_i\beta) = 0 \tag{4.4}$$

where $\widehat{W}_i = W(\widehat{\theta}; X_i)$. (Occasionally, when needed to stress the dependence of the estimator on the weights we will denote the solution of (4.4) by $\widehat{\beta}(\widehat{W})$.) The estimator $\widehat{\beta}$ is no longer necessarily unbiased because of the dependence of $\widehat{W}_i$ on the entire sample. However, under mild regularity conditions that include that $\widehat{\theta}$ has a probability limit θ^*, $W(\theta^*; X_i)$ is positive definite and $W(\theta; x)$ is a smooth function of θ and x, $\widehat{\beta}$ is a consistent estimator of β. Furthermore, $\sqrt{N}(\widehat{\beta} - \beta)$ converges as N goes to ∞ to a normal distribution with mean zero and variance C_W given by

$$C_W = \Gamma_W^{-1} \Omega_W \Gamma_W^{-1} \tag{4.5}$$

where

$$\Gamma_W = E\left\{X_i^T W\left(\theta^*; X_i\right) X_i\right\}$$

and

$$\Omega_W = E\left\{X_i^T W\left(\theta^*; X_i\right) \Sigma_i W\left(\theta^*; X_i\right) X_i\right\}.$$

See, for example, Newey and McFadden (1994), Theorems 6.1 and 6.2. The last result implies, in particular, that if $\widehat{\beta}(W)$ denotes the solution of (4.1) using $W_i = W(\theta^*; X_i)$, then the estimators $\widehat{\beta}(\widehat{W})$ and $\widehat{\beta}(W)$, appropriately normalized, have the same asymptotic distribution. This has the following interesting consequence. By (4.3), we have that

$$\operatorname{var}\left[\sqrt{N}\left\{\widehat{\beta}\left(\Sigma\right) - \beta\right\}\right] \leq \operatorname{var}\left[\sqrt{N}\left\{\widehat{\beta}\left(W\right) - \beta\right\}\right]$$

or equivalently,

$$\left(\frac{1}{N}\sum_{i=1}^{N} X_i^T \Sigma_i^{-1} X_i\right)^{-1}$$
$$\leq \left(\frac{1}{N}\sum_{i=1}^{N} X_i^T W_i X_i\right)^{-1} \left(\frac{1}{N}\sum_{i=1}^{N} X_i^T W_i \Sigma_i W_i X_i\right) \left(\frac{1}{N}\sum_{i=1}^{N} X_i^T W_i X_i\right)^{-1}.$$

Now when $W_i = \Sigma_i^{-1}$, $C_{\Sigma_i^{-1}} = E\{X_i^T \Sigma^{-1} X_i\}$. Thus, taking probability limit as N goes to infinity on both sides of the last inequality, and invoking the Law of Large Numbers we obtain

$$C_{\Sigma^{-1}} \leq C_W.$$

We conclude that the choice of weights $W_i = \Sigma_i^{-1}$ gives an asymptotically most efficient weighted least squares estimator, i.e., an estimator with asymptotic variance no greater than the asymptotic variance of any weighted least squares estimator. Furthermore, since $\widehat{\beta}(\widehat{W})$ and $\widehat{\beta}(W)$ have the same asymptotic distribution, then the estimator that uses as weights a consistent estimator of $\operatorname{var}(Y_i)$ is also an asymptotically efficient weighted least squares estimator.

This theory insures that for sufficiently large N, the effect of estimating the weight parameter θ is negligible. In particular, the weighted least squares estimate of β has the same asymptotic variance regardless of whether or not the weights are estimated or they are fixed known constants. However, in practice, if N is not too large, estimation of the weights may have an important effect on the actual variance of $N^{1/2}(\widehat{\beta} - \beta)$. Many theoretical studies for the case of Y_i a scalar outcome support this assertion. See, for example, Freedman and Peters (1984),

Carroll, Wu and Ruppert (1988), and reference therein. Carroll *et al.* (1988) examined in detail a special case in which $W(\theta; X_i)$ is the inverse of a correctly specified model for the variance function, $W(\theta; X_i)$ that does not depend on the mean, the errors have a symmetric distribution and θ is estimated by maximum likelihood assuming normal errors. They showed that in this case, $\widehat{C}_W$ is an underestimate of the actual variance of $N^{1/2}(\widehat{\beta} - \beta)$ if the kurtosis of the error distribution is less than or equal 2 but it is an overestimate otherwise. This is in accordance to the results of Freedman and Peters (1984) who noted the decrease in variance for heavy tail error distributions. Carroll *et al.* (1988) showed that the bootstrap variance estimate provides a more refined approximation to the actual variance of $N^{1/2}(\widehat{\beta} - \beta)$.

REMARK 1. Notice that the ML estimate takes exactly the same form as $\widehat{\beta}(\widehat{W})$, where $\widehat{\theta} = \widehat{\theta}_{\mathrm{ML}}$, and $\mathrm{var}(Y_i) = W_i(\widehat{\theta}, X_i)^{-1}$. Hence $\widehat{\beta}_{\mathrm{ML}}$ is asymptotically optimal, provided $\mathrm{var}(Y_i)$ is correctly specified. Otherwise $\widehat{\beta}_{\mathrm{ML}}$ is an ordinary WLS estimate with estimated weight matrix, and the asymptotic normal distribution previously given holds, i.e.,

$$\sqrt{N}\left(\widehat{\beta}_{\mathrm{ML}} - \beta\right) \sim N\left(0, C_W\right),$$

where C_W is given in (4.5).

REMARK 2. If the conditions given in Kackar and Harville (1981) hold, i.e., the variance-covariance component estimators are even and translation invariant and the error distributions for $(Y_i - X_i\beta)$ are symmetric, then WLS estimates with data-dependent weight functions are also unbiased for β in small samples.

4.4 Estimation of the Optimal Weight Function

Suppose we have complete data and $\Sigma_i = \Sigma$. A consistent estimator of the optimal weight Σ^{-1} can be obtained as follows. Suppose that $\widetilde{\beta} = \widehat{\beta}(W)$ is a weighted least squares estimator of β for an arbitrary (but fixed and known) choice of weight function. For example, $\widetilde{\beta}$ is the ordinary least squares estimator obtained from the choice $W_i = I$. Then, under mild regularity conditions on the weight function

$$\widehat{\Sigma} = S = \frac{1}{N}\sum_{i=1}^{N}\left(Y_i - X_i\widetilde{\beta}\right)\left(Y_i - X_i\widetilde{\beta}\right)^T \tag{4.6}$$

is a consistent estimator of Σ (see, e.g., Newey and McFadden, 1994, Lemma 4.3). Note that Σ is a one-step EM estimate, away from $\widehat{\Sigma}^0 = I$. See equation (3.11) and following.

The estimator $\widehat{\Sigma}$ separately estimates $n(n-1)/2$ non-diagonal and n diagonal elements. If n is large then a large sample size N is typically required for the asymptotic distribution of $\widehat{\beta}(\widehat{\Sigma})$ to be a good approximation of its finite sample distribution, and $\widehat{\beta}(\widehat{\Sigma})$ may have poor small sample behavior. Asymptotically efficient weighted least squares estimators with better small sample properties can be obtained under a parsimonious model for $\text{cov}(Y_i)$, i.e., when Σ is a known function $\Sigma(\theta)$ of a parameter vector θ of dimension $q \ll n(n+1)/2$. We now describe consistent estimators of θ under other models $\Sigma(\theta)$ that were introduced in Section 1.3. In what follows $\widetilde{\beta}$ is the weighted least squares estimator calculated using an arbitrary, but fixed and known, weight function.

Compound Symmetry. Recall from Section 1.3, Example 1, that Σ has a compound symmetry structure when $\Sigma = \sigma_1^2 I + \kappa 1 1^T$. Thus, letting $\theta_1 = \sigma_1^2 + \kappa$ and $\theta_2 = \kappa$, $\theta = (\theta_1, \theta_2)$, $\Sigma = \Sigma(\theta)$ is a symmetric matrix with all its diagonal elements equal to θ_1 and all its non-diagonal elements equal to κ. A consistent estimator of θ_1 is obtained by averaging the diagonal elements of $\widehat{\Sigma}$ defined in (4.6). Similarly, a consistent estimator of κ is obtained by averaging its non-diagonal elements. Thus,

$$\widehat{\theta}_1 = \frac{1}{n}\left[\sum_{j=1}^{n} S_{jj}\right]$$

and

$$\widehat{\theta}_2 = \frac{2}{n(n-1)}\left[\sum_{1\le j<k\le n} S_{jk}\right],$$

where S_{jj} and S_{jk} are elements of S defined in (4.6).

Banded. Recall from Section 1.3 that Σ has a banded structure if its (j,k)th and (j',k')th entries are equal when $j'-j = k'-k$. Thus, here $\Sigma = \Sigma(\theta)$ where $\theta = (\theta_1, \theta_2, \ldots, \theta_n)$ are the entries of Σ that vary freely. Using the notation of Section 1.3, $\theta_1 = \sigma^2$ and $\theta_{l+1} = \sigma^2 \rho_l$, $1 \le l \le n-1$. These can be consistently estimated by averaging the elements of $\widehat{\Sigma}$ defined in (4.6) over the entries known to be identical under the banded structure. Thus, for example, $\widehat{\theta}_1$ is the same as for the compound symmetry model, while for $1 \le l \le n-1$,

$$\widehat{\theta}_{l+1} = \frac{2\widehat{\theta}_1}{(n-l)}\left[\sum_{k=1}^{n-l}\left(S_{k(l+k)}\right)\right] \tag{4.7}$$

REMARK. Maximum likelihood or REML estimates of θ will also be consistent for θ in this setting, even without the assumption of normality. They can be computed as described in 3.

4.5 Locally Optimal Weighted Least Squares

Once the estimator $\Sigma(\widehat{\theta})$ of Σ is obtained, estimation of β proceeds by solving the weighted least squares equation using the weight matrices $\Sigma_i(\widehat{\theta})^{-1}$. The estimation procedure is then a three stage procedure summarized as follows:

1. Obtain a preliminary weighted least squares estimator $\widetilde{\beta}$ (e.g., the ordinary least squares estimator),
2. Estimate Σ with any consistent estimate $\widehat{\Sigma}$ (as defined in (4.6) for example) or with $\widehat{\Sigma}_i = \Sigma_i(\widehat{\theta})$ where $\widehat{\theta}$ is a consistent estimator of θ under a model for $\text{cov}(Y_i)$,
3. Obtain the solution $\widehat{\beta}(\widehat{\Sigma})$ to the weighted least squares estimating equation (4.4) with weights $\widehat{W}_i = \widehat{\Sigma}_i$.

The estimator $\widehat{\beta}(\widehat{\Sigma})$ is consistent and asymptotically normal. We say that this estimator is *a locally optimal weighted least squares* (*LOWLS*) *estimator* because it has asymptotic variance that is equal to the lower bound for the variances of all weighted least squares estimators when the model for $\text{cov}(Y_i)$ is correctly specified and $\widehat{\beta}(\widehat{\Sigma})$ remains consistent and asymptotically normal even if the model for $\text{cov}(Y_i)$ is misspecified. Notice that the LOWLS makes no reference to the distribution of the error terms. In general, $\widehat{\beta}(\widehat{\Sigma})$ will be close to the ML estimate of β obtained by assuming normality, and the ML estimate has the same properties in this setting.

4.6 Model Based and Robust Variance Estimation

In this section we consider estimation of the asymptotic variance of $\widehat{\beta}(\widehat{\Sigma}^{-1})$. By the results of Section 4.3, $\sqrt{N}\{\widehat{\beta}(\widehat{\Sigma}^{-1})-\beta\}$ has an asymptotic mean zero normal distribution with variance equal to $C_{\Sigma^{-1}}$ provided $\widehat{\Sigma}$ is a consistent estimator of Σ, i.e., provided the model used to estimate

$\text{cov}(Y_i)$ is correctly specified. When this is the case, a consistent estimator of $C_{\Sigma^{-1}}$, and therefore of the asymptotic variance of $\sqrt{N}\{\widehat{\beta}(\widehat{\Sigma}^{-1})-\beta\}$ is given by

$$\frac{1}{N}\sum_{i=1}^{N} X_i^T \widehat{\Sigma}^{-1} X_i. \tag{4.8}$$

If the model for $\text{cov}(Y_i)$ was incorrectly specified, then by the results of Section 2.3, $\sqrt{N}\{\widehat{\beta}(\widehat{\Sigma}^{-1}) - \beta\}$ has the same asymptotic variance as $\sqrt{N}\{\widehat{\beta}(\Sigma^{*^{-1}})-\beta\}$ where Σ^* is the probability limit of $\widehat{\Sigma}$. This asymptotic variance is equal to

$$C_{\Sigma^{*-1}} = \Gamma^{-1}_{\Sigma^{*-1}} \Omega_{\Sigma^{*-1}} \Gamma^{-1}_{\Sigma^{*-1}}.$$

A consistent estimator of $\Gamma_{\Sigma^{*-1}}$ is given by

$$\widehat{\Gamma}_{\Sigma^{*-1}} = \frac{1}{N}\sum_{i=1}^{N} X_i^T \widehat{\Sigma}^{-1} X_i,$$

and a consistent estimator of $\Omega_{\Sigma^{*-1}}$ is given by

$$\widehat{\Omega}_{\Sigma^{*-1}} = \frac{1}{N}\sum_{i=1}^{N} X_i^T \widehat{\Sigma}^{-1} \left(Y_i - X_i\widehat{\beta}\right)\left(Y_i - X_i\widehat{\beta}\right)^T \widehat{\Sigma}^{-1} X_i.$$

Thus, a consistent estimator of $C_{\Sigma^{*-1}}$ is given by

$$\widehat{\Gamma}^{-1}_{\Sigma^{*-1}} \widehat{\Omega}_{\Sigma^{*-1}} \widehat{\Gamma}^{-1}_{\Sigma^{*-1}}. \tag{4.9}$$

The variance estimator (4.8) is often referred to as a *model based* variance estimator to remind the user that its consistency relies on the correct specification of the model for $\text{cov}\,(Y_i)$. The variance estimator (4.9) is often referred to as a *robust* variance estimator since it is consistent even under misspecification of the model for $\text{cov}(Y_i)$. It has also been referred to as the sandwich or empirical variance estimator (Liang and Zeger, 1986). The usual trade off between bias and variance applies to these two choices. The sandwich variance estimator is asymptotically unbiased regardless of model specification. However, even with large samples it will typically have bigger fluctuations around the true asymptotic variance than the model based variance estimator unless the assumed model is far from the true variance. Also, under correct model specification the model based variance estimator will typically have smaller mean squared error than the robust variance estimator.

The variance estimators can be used to construct Wald tests of constraint null hypotheses of the form

$$H_0 : \underset{q\times p}{Q}^T \underset{p\times 1}{\beta} = 0.$$

Specifically, let $\widehat{\beta}$ denote $\widehat{\beta}\left(\widehat{\Sigma}^{-1}\right)$. Under the null hypothesis, we have

$$\sqrt{N} Q^T \widehat{\beta} \xrightarrow{\mathcal{L}} N_q\left(0; Q^T C_{\Sigma^{*-1}} Q\right)$$

then

$$\widehat{\beta}^T Q \left(Q^T \widehat{C}_{\Sigma^{*-1}} Q\right)^{-1} Q^T \widehat{\beta} \tag{4.10}$$

has an asymptotic χ^2_q distribution. A test that rejects when (4.10) exceeds the α point of the χ^2_q distribution is a valid asymptotic α level test of H_0.

REMARK. The results in this section can be applied straightforwardly to obtain $\text{var}(\widehat{\beta}_{\text{ML}})$ when the form of $\text{var}(Y_i)$ is unknown, and some user-specified model has been selected for $\Sigma_i(\theta)$.

4.7 Joint Estimation of β and θ

The development in Sections 4.1-4.6 was aimed at inference about the parameter β. In Section 4.4 the emphasis was on estimation of $\text{cov}(Y_i)$ for improvement of the efficiency of the weighted least squares estimator of β. Notice, for example, that although easy to obtain, in the cases described in Section 4.4, we have not given formulas for consistent estimators of the variance of $\widehat{\theta}$. In fact, nowhere in the derivation of consistency and asymptotic normality of the weighted least squares estimators of β is it required that a model for cov (Y_i) be specified. That is, the WLS estimators are consistent and asymptotically normal (CAN) under the sole assumption that the conditional mean of Y_i is linear in the covariates, i.e., that

$$E\left(Y_i\right) = X_i^T \beta. \tag{4.11}$$

This has the advantage that inferences about β are robust to misspecification of the model for $\text{cov}(Y_i)$. However, even if inference about β was the sole primary goal of the analysis, the WLS approach described above has the drawback that under most error distributions, knowledge of the model for $\text{cov}(Y_i)$, when available, can be used to improve the efficiency with which β is estimated. We can obtain efficiency improvements over the weighted least squares estimators described above by further enlarging the class of estimating equations and calculating the optimal estimator under the larger class. This can be effectively done for many distributions, by solving joint estimating equations for β and θ. In addition, as a by-product of jointly estimating β and θ, consistent variance estimator formulae for θ can be readily derived from the standard theory of estimating equations, therefore providing the possibility of conducting inference about θ as well. We postpone the discussion of semi-parametric, joint estimation of the mean and second moment parameters to Chapter 6, where we will discuss the generalized linear multivariate model.

4.8 Efficiency of OLS Estimators

It is often suggested that using $W_i = I$ (or using OLS) leads to an estimator with nearly the same efficiency as the optimal estimate using Σ_i^{-1}, in many cases. The claim can be justified by considering simple forms for Σ, eg compound symmetry, or banded, with constant variance, and/or simple forms for each X_i. Bloomfield and Watson (1975) gave a general formula for relative efficiency of $\widehat{\beta}(I)$ and $\widehat{\beta}(\Sigma^{-1})$, by comparing

$$\left(\sum_{i=1}^{N}(X_i^T X_i)\right)^{-1}\left(\sum_{i=1}^{N} X_i^T \Sigma_i X_i\right)\left(\sum_{i=1}^{N}(X_i^T X_i)\right)^{-1}$$

with

$$\left(\sum_{i=1}^{N}\left(X_i^T \Sigma_i^{-1} X_i\right)\right)^{-1}.$$

We consider some examples given in Diggle *et al.* (1994).

Example 1.

$$N = 10 \; n = 5 \; t_j = (-2, -1, 0, 1, 2)$$

$$\mathrm{E}(Y_{ij}) = \beta_0 + \beta_1 t_j.$$

If there are no missing observations and $\Sigma = \sigma^2\left((1-\rho)I + \rho J\right)$, then $\widehat{\beta}(\Sigma^{-1}) = \widehat{\beta}(I)$ and OLS is optimal. This requires a common design, no missing data and compound symmetry. As the next example shows, efficiency of OLS in this case is still high with an autoregressive structure.

Example 2.

Suppose the design on time remains the same, but Σ is of the form

$$\Sigma = \sigma^2 \begin{bmatrix} 1 & \rho & \rho^2 & \rho^3 & \rho^4 \\ & 1 & \rho & \rho^2 & \rho^3 \\ & & 1 & \rho & \rho^2 \\ & & & 1 & \rho \\ & & & & 1 \end{bmatrix}$$

(still equal spacing and equal variances on diagonal). It is straightforward to show (Diggle *et al.*, 1994) that

$$\text{var } \widehat{\beta}_{\text{OLS}} = \begin{bmatrix} 10(5 + 8\rho + 6\rho^2 + 4\rho^3 + 2\rho^4) & 0 \\ 0 & 20(5 + 4\rho - \rho^2 - 4\rho^3 - 4\rho^4) \end{bmatrix}$$

and

$$\text{var } \widehat{\beta}\,(\Sigma^{-1}) = \sigma^2(1-\rho^2) \begin{bmatrix} 0.1(5 - 8\rho + 3\rho^2)^{-1} & 0 \\ 0 & 0.05(5 - 4\rho + \rho^2)^{-1} \end{bmatrix}.$$

The relative efficiencies can be obtained by looking at the ratio of the diagonals. They are computed for a range of ρ's below (taken from Diggle *et al.*, 1994):

ρ	0.1	0.2	0.3	0.4	0.5
$e(\beta_0)$	0.998	0.992	0.983	0.973	0.963
$e(\beta_1)$	0.997	0.989	0.980	0.970	0.962

ρ	0.6	0.7	0.8	0.9	0.99
$e(\beta_0)$	0.955	0.952	0.956	0.970	0.996
$e(\beta_1)$	0.952	0.955	0.952	0.955	0.961

These are all very close to 1, because $n_i = n$, $X_i = X$, and $\text{var}(Y_{ij})$ is constant.

Example 3.

A crossover design where carryover may be present (Fitzmaurice *et al.*, 1993). Here $n = 3$ and there are 2 treatments; subjects are assigned in equal numbers to all possible 2^3 sequences: AAA, AAB, ABA, ABB, BAA, BAB, BBA, BBB. Here

$$\text{E}(Y_{ij}) = \beta_0 + \beta_1 t_{ij}$$

where $t_{ij} = 1$ if ith subject gets A in the jth period. Note that there are eight distinct X_i's, one for each sequence. Assuming Σ has an autocorrelation structure, Diggle *et al.* (1994) obtain the following efficiency ratios:

ρ	0.1	0.2	0.3	0.4	0.5
$e(\beta_0)$	0.993	0.974	0.946	0.914	0.880
$e(\beta_1)$	0.987	0.947	0.883	0.797	0.692

ρ	0.6	0.7	0.8	0.9	0.99
$e(\beta_0)$	0.846	0.815	0.788	0.766	0.751
$e(\beta_1)$	0.571	0.438	0.297	0.150	0.015

The efficiency for estimating the treatment effect declines substantially for $\rho > 0.2$. An intuitive explanation for this is as follows. To estimate a treatment contrast, some individuals provide more information than others; eg., a person in sequence 1 and another in sequence 8 provide the contrast $y_{i1} - y_{i'1}$ to estimate $A - B$, with variance equal to $2\sigma^2$. Now consider within subject contrasts:

$$
\begin{array}{ll}
y_{i2} - y_{i3} & \text{for sequence 2} \\
y_{i1} - y_{i2} & \text{for sequence 3} \\
y_{i1} - y_{i2} & \text{for sequence 4} \\
\text{etc.} &
\end{array}
$$

Each has variance $2\sigma^2(1 - \rho)$, so if ρ is large, these within subject contrasts have *much more* information and should be weighted more heavily. Here correlation is important because of the design, variance is less so. The OLS ignores the design and weights all possible contrasts equally.

4.9 Remarks

The asymptotic properties of extremum estimators that minimize some objective function have been widely studied, among others by Fisher (1925), Wald (1949), Huber (1967), Jenrich (1969) and Amemiya (1973, 1985). White (1980) specialized these results to obtain robust variance estimators for the LOWLS estimators of Section 2.5. The LOWLS estimators are sometimes also referred to as the Generalized Estimating Equations (GEE) estimators after the papers of Liang and Zeger (1986) and Zeger and Liang (1986). These papers considered estimating equations for a more general class of mean models that include generalized linear models. These authors call the model for cov (Y_i) a "working covariance model" as it is indeed not needed for consistency and asymptotic normality of the LOWLS estimator. The GEE methodology will be taken up in Chapter 6.

The weighted least squares estimators are *semiparametric* estimators because they are CAN without requiring a full parametric description of the law of the data. In fact, they are CAN provided only that the model (4.11) for the conditional mean of the outcome vector is correctly specified. That is, consistency and asymptotic normality is obtained without specifying any restriction on the error distribution besides the conditional mean zero restriction.

4.10 Studies with Clusters of Random Size or Missing Data

In some designs, the cluster size may not be fixed in advance. Suppose for example N units are randomly drawn from a specific population. Each unit i, $i = 1, \ldots, N$, is composed of n_i members. On each member $j, j = 1, \ldots, n_i$, of each unit i, we observe an outcome Y_{ij} and an associated vector of covariates X_{ij}. The covariates X_{ij} may be cluster-specific, i.e., X_{ij} varies across units but not across members of the same unit so that X_{ij} is the same for all j, subject-specific, i.e., X_{ij} varies across members of the same unit, or both. Thus, $Y_i = (Y_{i1}, \ldots, Y_{in_i})'$ and $X_i = (X_{i1}, \ldots X_{in_i})'$ record the full data on the ith sampling unit. As another example, suppose we randomly sample N units from the population and intend to make n observations on each subject with $n \times p$ design matrix X_i^C. Suppose in fact we are able to obtain only n_i observations for each subject, but the missing observations are MCAR, so that for the observed data vector, $E(Y_i) = X_i\beta$, where the rows of X_i are the subset of rows of X_i^C corresponding to the observed Y_i.

The LMCD (1.1.2) specifies that the conditional mean of each observation Y_{ij} depends linearly on functions of cluster-specific and subject-specific covariates. It also specifies the functional dependence of the covariance between each pair of observations on covariates. Thus, Σ_i depends on i through its dimension $n_i \times n_i$, and through its postulated, if any, dependence on the covariate matrix X_i. Often the values of n_i are unknown to the investigator prior to the collection of data. In such settings, cluster size is also a random variable. It can be shown that the asymptotic results given for fixed n_i also hold provided we assume each $n_i <$ some fixed value n^*. Although it was not explicitly stated in section 2, when n_i is random, the LMCD is a model for the conditional mean and covariance of Y_i given the covariates X_i and the cluster size n_i. Notice, however, that as formulated, the LMCD does not allow for the mean and covariance of Y_{ij} to depend on cluster size other than through dimensionality, nor for covariate effects to change with cluster size. It is therefore inappropriate for analyzing studies where such relationships are of interest. An illustration of this scenario would be a study of birth weight in animals in which the primary sampling units are litters and the effect of litter size on birth weight is of scientific interest. The LMCD can be easily extended to allow for the dependence of the mean and covariance of Y_i on n_i beyond dimensionality but such extension will not be elaborated in this monograph.

Because the WLS estimate is only consistent when $E(Y_i) = X_i\beta$, when the lack of balance is due to missingness, an MCAR mechanism is required for the validity of the WLS estimate. Because $\widehat{\beta}_{\text{ML}}$ is a WLS estimate, it too is automatically consistent for β even if the error distributions are not normal, provided we have $E(Y_i) = X_i\beta$. The same is true for ML or REML estimates of θ, i.e., normality of the error distributions is not required for consistency. With missingness due to an MAR mechanism, then it generally no longer holds that $E(Y_i) = X_i\beta$, and WLS estimates are no longer valid. ML estimates for both β and θ are consistent in the MAR setting, although now the distributional assumption of the error terms is now crucial, as is the assumption that $\text{var}(Y_i) = \Sigma_i$ is correctly specified.

Chapter 5

Random Effects and the Linear Mixed Model

Until now, we have considered primarily the estimation of regression parameters from the linear model, i.e., β in the model

$$E\ (Y_i|X_i) = X_i\beta, \tag{5.1}$$

as well as the variance parameters when $\text{var}(Y_i)$ has an arbitrary structure. In this chapter we consider the use of random effects in modeling longitudinal data or clustered outcomes. Many researchers view it as more natural to assume that the mean response depends upon a combination of population parameters β and subject-specific effects. In the setting of the linear model where Y_i is linear in the parameters and the error terms, it is natural to also assume Y_i is linear in the subject-specific effects. As we will show, this still leads to the linear model (5.1), but $\text{var}(Y_i)$ now has a special random effects structure.

The use of random effects offers several benefits when modeling longitudinal data. First, it provides a way to model correlation in unbalanced designs. Secondly, random effects can be used to estimate subject-specific effects arising in several applications. Finally, it offers an optimal way to combine within- and between-subject data.

Random effects are useful when strict measurement protocols are not followed and we have measurements made at arbitrary, irregularly spaced intervals. It is not desirable to design a study in this way, but such data sets are not uncommon. It can happen that we start with a strict protocol but because of missingness and missed timing, we end up with measurement times that do not conform to a set of protocols-defined occasions. Use of retrospectively collected records for analysis also often leads to unbalanced designs. Random effects models handle this in a very natural way. The same is true with clustered data, where units are nested within cluster. Another setting is where the metameter chosen for analysis differs from age or time, as in using current height to predict lung function, i.e., should variance in lung function depend on age, occasion of measurement, height or some combination? Finally, we may choose to rescale the time variable differently for each subject, to reflect time before and after a critical event, such as menarchy or sero-conversion.

This chapter is organized as follows. First we introduce ideas in terms of a two-stage random effects model with both population parameters and subject effects, then we consider general linear mixed model, estimation of β and Σ by ML and REML estimation, and finally estimation of the random effects.

5.1 Two-Stage Random Effects Models

Two-stage random effects models begin by assuming at Stage 1 that each unit has its own design on time, denoted by Z_i, and its own parameter vector β_i. Given Z_i and β_i, we assume that:

Stage 1.

$$\underset{n_i\times 1}{Y_i} = \underset{n_i\times q}{Z_i}\ \underset{q\times 1}{\beta_i} + \underset{q\times 1}{e_i} \tag{5.2}$$

where $e_i \sim N_{n_i}(0, \sigma^2 I)$, and the e_i are independent. The e_{ij}'s are *iid* $N(0, \sigma^2)$, so can be thought of as measurement error. The β_i are often called the "true regression coefficients," since the observed responses for the ith subject are assumed to follow the curve with coefficients β_i, but with added measurement error e_i. This defines Stage 1 of the model.

Here Z_i specifies the growth curve model, such as linear or quadratic (or a spline, etc.) or more generally, the pure "within subject covariates." For example, we may have

$$Y_i = \begin{pmatrix} 1 & t_{i1} \\ 1 & t_{i2} \\ \vdots & \vdots \\ \vdots & \vdots \\ 1 & t_{in_i} \end{pmatrix} \begin{pmatrix} \beta_{0i} \\ \beta_{1i} \end{pmatrix} + \begin{pmatrix} e_{i1} \\ \vdots \\ \vdots \\ e_{in_i} \end{pmatrix}. \tag{5.3}$$

Notice that q, the dimension of β_i, does not vary with i. Apart from the intercept, Z_i contains only within subject covariates. We accommodate subject-specific covariates in Stage 2. A way to think about the Stage 1 model is that generally we want to be able to obtain estimates of β_i and σ^2 using just the data from the ith subject, i.e.,

$$\widehat{\beta}_i = \left(Z_i^T Z_i\right)^{-1} Z_i^T Y_i,$$

$$\widehat{\sigma}^2 = \sum_{i=1}^{N} Y_i^T \left(I - Z_i \left(Z_i^T Z_i\right)^{-1} Z_i^T\right) Y_i \Big/ \sum_{i=1}^{N} (n_i - q)$$

This may not be possible for each subject (if, e.g., $n_i \leq q$), but it should be feasible in principle. Thus between subject variables, such as sex, cannot be included at Stage 1. Essentially, Stage 1 consists of separate regression models for each subject, with the same set of predictors, but possibly different values for the predictors.

REMARK. Note that we could allow correlation between $e_{ij}, e_{ij'}$, but this would change the interpretation of the β_i's. Assuming correlation between the error terms implies that e_{ij} is no longer simply "measurement error" but includes model misspecification at the individual level. Alternately, we can have two "error" components, one modeling serial correlation resulting from model misspecification, and one modeling measurement error, as in the Diggle (1988) model discussed in Chapter 1.

The β_i's are random variables; to specify population parameters we model the mean and variance of the random effects at Stage 2. We model variation in the β_i's as a function of subject-specific covariates and residual between subject variation:

Stage 2.

$$\underset{q\times 1}{E\left(\beta_i\right)} = \underset{q\times p}{A_i} \; \underset{p\times 1}{\beta}, \tag{5.4}$$

$$\mathrm{var}(\beta_i) = \underset{q\times q}{D}. \tag{5.5}$$

This completes the Stage 2 model. This model allows some of the variation in the β_i's to be explained by covariates contained in A_i; the remaining variation is measured by D. Note that A_i cannot include within subject covariates, such as time, because the outcome of the model is "β_i" which does not vary over time.

EXAMPLE. Suppose we have two groups, and we model a linear decline in each, with mean slope and intercept depending upon group, i.e., we use model (5.3) to describe individual responses, Y_{ij}. Here t_{ij} is the time (since beginning of study) that the jth measurement was made for the ith subject. In other settings, t_{ij} might be age of subject. Denote the group by a dummy variable:

$$\begin{aligned} g_i &= 1 \quad \text{if GR 1,} \\ g_i &= 0 \quad \text{if GR 0.} \end{aligned}$$

Consider a model that assumes the mean slope and intercept differs from each group:

$$\begin{aligned} E(\beta_{0i}) &= \beta_1 + \beta_2 g_i, \\ E(\beta_{1i}) &= \beta_3 + \beta_4 g_i. \end{aligned} \tag{5.6}$$

Here β_1 is the the mean intercept for group 0, and $\beta_1 + \beta_2$ is the mean intercept for group 1, so β_2 is the difference in the intercepts in the two groups; β_3 is the mean slope in group 0 and β_4 is the difference on the slopes. Thus β_2 might be considered the main effect of group and β_4 the group $\times$ time interaction effect, where the time trend is assumed to be linear.

In matrix notation, (5.6) implies

$$E(\beta_i) = A_i \begin{pmatrix} \beta_1 \\ \beta_2 \\ \beta_3 \\ \beta_4 \end{pmatrix}$$

where

$$A_i = \begin{bmatrix} 1 & g_i & 0 & 0 \\ 0 & 0 & 1 & g_i \end{bmatrix}.$$

Further assume that regardless of treatment group,

$$\text{var}\,(\beta_i) = D = \begin{bmatrix} d_{00} & d_{01} \\ d_{10} & d_{11} \end{bmatrix}$$

where

$$\begin{aligned} d_{00} &= \text{var}(\beta_{0i}), \\ d_{11} &= \text{var}(\beta_{1i}), \\ d_{10} &= \text{cov}(\beta_{0i}, \beta_{1i}). \end{aligned}$$

The variation in β_{0i} after adjusting for person specific covariates is d_{00}, and similarly for d_{11}. Figure 5.1 illustrates a scenario for group 0, where individual curves are plotted using $\beta_{0i} + \beta_{1i}t$. The heavy line represents the curve obtained from plotting $\beta_1 + \beta_3 t$ for group 0.

In this picture there is substantial variability in the intercepts (d_{00}), but apart from two subjects, the slopes are nearly constant ($d_{11} \doteq 0$).

REMARK. The coding of the Stage 1 regression variables is critical for the proper interpretation of the mean and variance effects. Returning to

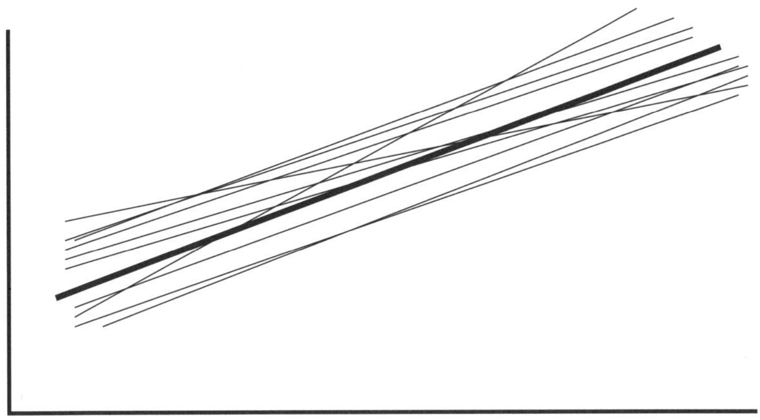

FIGURE 5.1.

our example, if t_{ij} is time since baseline (so $t = 0$ denotes baseline), then β_{0i} represents the expected response at baseline. If t_{ij} is age (say a_{ij}), then the β_{0i}'s will rarely have a useful interpretation because they represent expected response at age zero. In this setting one could contemplate coding t_{ij} as $(a_{ij} - \overline{a}_{i\cdot})$ or $(a_{ij} - \overline{a}_{\cdot\cdot})$ or $(a_{ij} - A)$, where A is some fixed age in the age range of participants. This last choice will sometimes be most attractive since it makes the β_{0i}'s interpretable as the individual's expected response at the common age A, and d_{00} is the variation around the response at age A. Notice that a response at A *need not be* observed for any subject as the model is used to predict the response at A for each subject. If $t_{ij} = (a_{ij} - \overline{a}_{i\cdot})$ then β_{0i} estimates the subject's response at their mean age over the period of follow-up. This may vary considerably from subject to subject, and create a large value for d_{00} which is not meaningful. This is rarely a good choice for centering.

In some circumstances, it may be useful to use $t_{ij} = (a_{ij} - a_i^*)$, where a_i^* is the age that the ith individual experiences a benchmark event. Namova *et al.* (2001) use random effects models to characterize change in body fat in girls before and after menarchy. The study was designed to begin annual follow-up prior to menarchy and continue for four years after menarchy. For analysis, time is coded as time since menarchy, and could be positive or negative. The intercepts, β_{0i}'s, provide an estimate of the individuals body fat at menarchy, even though no actual measures of body fat were made at this occasion, except by chance.

Returning to the two-stage model, we use the identities

$$E(Y_i|X_i) = E(E(Y_i|Z_i, \beta_i))$$

and
$$\text{var}(Y_i|X_i) = E\ \text{var}(Y_i|Z_i, \beta_i) + \text{var}E(Y_i|Z_i, \beta_i)$$

where X_i contains the covariates in Z_i and A_i, to show that

$$E(Y_i|X_i) = E(Z_i\beta_i) = Z_iA_i\beta$$

and

$$\text{var}(Y_i|X_i) = Z_iDZ_i^T + \sigma^2 I. \qquad (5.7)$$

Hence we have the same general linear model for correlated data, except:

1. We have a special structure for X_i given by $X_i = Z_iA_i$, which implies also that all covariates can all be classified as either "within" or "between" subject covariates. As discussed in the next section, this can be a serious limitation.

2. We have a special random-effects structure for $\text{var}(Y_i) = \Sigma_i$ as $Z_iDZ_i^T + \sigma^2 I$. Notice that this allows variances and covariances to depend on individual times of measurements. Notice also that the number of variance-covariance parameters does not depend upon the n_i.

Returning to the example, where $Z_{ij} = (1, t_{ij})$, (5.7) implies that

$$\begin{aligned} \text{var}\,(Y_{ij}) &= (1, t_{ij})\, D \begin{pmatrix} 1 \\ t_{ij} \end{pmatrix} + \sigma^2 \\ &= d_{00} + 2t_{ij}d_{10} + t_{ij}^2 d_{11} + \sigma^2 \end{aligned}$$

depends upon t_{ij}, it is easily shown that $\text{cov}\left(Y_{ij},\, Y_{ij'}\right)$ depends upon $t_{ij}, t_{ij'}$. Hence $\Sigma_i(\theta)$ depends upon the pattern of observation times, as well as the variance and covariance parameters. Notice that only four parameters ($\sigma^2, d_{00}, d_{10}, d_{11}$, or in general, $1 + q(q+1)/2$) are needed to model $\text{var}(Y_i)$.

REMARK. The case where $n_i = n$, and $Z_i = Z$, is very special. Here each subject has exactly the same design on time. In this setting it is well known that in the absence of subject-specific covariates, the population mean curve is the same as the average of the individual curves. In addition, OLS estimates of each subject have identical precision (assuming σ^2 is the same for each subject).

If in addition, A_i can be formulated as $A_i = a_i^T \otimes I$, where a_i is a $k \times 1$ vector of subject-specific covariates and I is a $q \times q$ identity matrix, then the regression of each component of β_i on the covariates has the same design matrix; i.e., for some partition of β we can write the following:

$$\begin{aligned} E(\beta_{0i}) &= a_i^T \beta^0 \\ E(\beta_{1i}) &= a_i^T \beta^1 \\ &\vdots \\ E(\beta_{qi}) &= a_i^T \beta^q, \end{aligned}$$

where the β^0, β^1, ..., β^q are distinct vectors, each of length $k \times 1$, so that β is $gk \times 1$. Essentially $A_i = a_i^T \otimes I$ means that the covariate models are the same for each random effect, i.e., if sex affects the intercept, it affects the slope, etc., as well.

In this special setting, one can sometimes reduce the estimation problem to a series of univariate regressions without any loss of efficiency. First, each β_i is estimated by OLS for each subject, then each element of $\widehat{\beta}_i$ is analyzed using univariate regression methods. This approach is fully efficient if the assumption that $\Sigma = Z\,D\,Z^T + \sigma^2 I$ holds, and it may be reasonably efficient under weaker assumptions. Of course, one cannot use univariate methods to test for global covariate effects on time, and there will be further loss of efficiency with unbalanced designs of missing data, or settings where the design matrix departs from $a_i^T \otimes I$. But it does suggest that this simple two-step strategy is worth considering if the design is nearly balanced, and the covariates affect all of the model coefficients.

5.2 A Linear Mixed Model (LMM)

In this section we reformulate the two-stage random effects model as a more general mixed model that allows greater flexibility in handling all types of covariates. We rewrite β_i as

$$\beta_i = A_i\beta + b_i$$

where $b_i \sim N_q(0, D)$. Hence b_i gives the coefficients for an individual's residual curve after the covariate effects have been accounted for. For the example in 5.1, b_{0i} is an individual's deviation from the mean intercept for their treatment group, and b_{1i} is their residual from the mean slope.

We can now combine Stages 1 and 2 of the random effects model to write

$$\begin{aligned} Y_i &= Z_i\,(A_i\beta + b_i) + e_i \\ &= (Z_iA_i)\,\beta + Z_ib_i + e_i. \end{aligned}$$

Thus we have partitioned Y_i into three components:

$$Y_i = \text{mean} + \begin{array}{c}\text{“between subject}\\ \text{residual”}\end{array} + \begin{array}{c}\text{“within subject}\\ \text{residual”}\end{array}.$$

The between subject residual is b_i with zero mean and variance D, and e_i is the within subject residual with zero mean and variance $\sigma^2 I_{n_i}$. It follows that

$$E(Y_i) = (Z_i A_i)\beta$$

and

$$\text{var}\,(Y_i) = Z_i D Z_i^T + \sigma^2 I.$$

Notice that this two-stage derivation requires that $X_i = Z_i A_i$ have a special structure which is inconvenient, since A_i must have only non-time-varying covariates and Z_i has only time-varying covariates (except the intercept). In order to allow for a sufficiently complex structure for the mean response $(Z_i A_i)\beta$, it may be necessary to include many variables in Z_i, requiring an equally complex Σ. Note that the same Z_i appears in both $E(Y_i)$ and var(Y_i). Simply modifying the model to allow X_i to include whatever covariates we want without changing Z_i gives more flexibility. For example, in modeling mean lung function in children as linear in age and height, using the two-stage model would require D to be at least 3×3 since both age and height vary over time within a subject.

Once we get away from the concept of two-stage, not only can we let X_i be arbitrary, we can also choose q arbitrarily, i.e., we could assume the intercepts vary randomly but the slopes do not, in which case Z_i is just an $n_i \times 1$ vector of ones. This yields compound symmetry:

$$\begin{aligned}\text{var}(y_{ij}) &= d_{00} + \sigma^2,\\ \text{cov}(y_{ij},\, y_{ij'}) &= d_{00}.\end{aligned}$$

Thus we define the linear mixed model (LMM) as follows:

$$\boxed{\underset{n_i\times 1}{Y_i} = \underset{n_i\times p}{X_i}\;\underset{p\times 1}{\beta} + \underset{n_i\times q}{Z_i}\;\underset{q\times 1}{b_i} + \underset{n_i\times 1}{e_i}\,.} \qquad (5.8)$$

Both b_i and e_i are independent, zero mean error terms. The only constraint on X_i and Z_i is that Z_i be a subset of the columns of X_i. This is because we think of $Z_i\, b_i$ as zero mean residuals, so they should “deviate” from some corresponding nonzero mean. For example, it would be counterintuitive to let the individual slopes deviate in the population and assume that the population mean slope is zero.

Notice that (5.8) is like the LMCD for correlated data in that

$$E\left(Y_{n_i\times 1}\right) = X_i\beta$$

but now

$$\text{var}\left(Y_{n_i\times 1}\right) = Z_iDZ_i^T + \sigma^2 I$$

has the characteristic random effects structure. We also have random effects (b_i's) which can be estimated. We can generalize this model even further to allow var$(e_i) = R_i$. Notice that R_i cannot be completely unstructured because we will have overparameterized. Generally, both the dimension and the elements of R_i may depend on unique times of measurement.

Sometimes, as in the Diggle (1988) model, R_i is expanded further to include both a serial measurement component, say S_i and a measurement, or sampling error component, say M_i, with var$(M_i) = \sigma^2 I_{n_i\times n_i}$ and var $S_i = \tau^2\Omega_i$; Ω_i may have one of the structures considered in Section 1.4. Hence $R_i = \sigma^2 I + \tau^2\Omega_i$. As noted above, for arbitrary Z_i, D and Ω_i, there are identifiability problems which are not readily quantified for the general case, hence the different components of

$$\text{var}\left(Y_i\right) = Z_iDZ_i^T + R_i$$

should be kept relatively simple. McCulloch and Searle (2001) discuss several models for var(Y_i) in this setting.

5.3 ML Estimation for the LMM

For the purpose of estimating β and θ, where θ is the vector of parameters in D and R_i, the LMM can be viewed as a special case of the LMCD where Σ_i has a variance component structure. Notice that specification of the LMM does not require normally distributed error terms. When Σ_i is known, the optimal estimator (and also the ML assuming normality) of β is $\widehat{\beta}(\Sigma_i^{-1})$. Method-of-moment, ML and REML estimates for θ have been proposed (Jennrich and Schluchter, 1986; Laird and Ware, 1984; Vonesh and Carter, 1982). We will discuss only ML and REML estimation; software is readily available for ML and REML. As in the general case, ML and REML esimates are consistent even in the absence of normality. In some settings we may also want to estimate the individual random effects (b_i's). A special theory is needed for these; it will be discussed in 5.5. For the remainder of this section we assume that $b_i \sim N_q(0, D)$, $e_i \sim N_{n_i}(0, R_i)$ and given X_i, Z_i the Y_i's are independent, $i = 1, \ldots, N$ so that we may derive ML and REML estimates of θ.

First consider estimating $\widehat{\theta}$ via ML (assuming $R_i = \sigma^2 I$). By definition $(\widehat{\beta}_{ML}, \widehat{\theta}_{ML})$ are obtained by maximizing $\mathcal{L}(\beta, \theta)$ over (β, θ) where

$$\mathcal{L}(\beta, \theta) = \prod_{i=1}^{N} \frac{1}{|\Sigma_i|^{1/2}} e^{-1/2(Y_i - X_i\beta)^T \Sigma_i^{-1}(Y_i - X_i\beta)}$$

$$\Sigma_i = Z_i D Z_i^T + \sigma^2 I,$$

and

$$\theta^T = (\sigma^2,\ d_{00},\ d_{11},\ d_{10},\ \text{etc.}).$$

As before, $\widehat{\beta}(\Sigma_i^{-1})$ can be obtained in closed form for given θ. While expressions for the likelihood equations for θ can be found by directly differentiating the likelihood and substituting in $\widehat{\beta}(\Sigma_i^{-1})$ for β, it is instructive, and easier, to use the EM approach to derive them, even though most computing packages do not routinely use the EM for the computations.

Here it is most convenient to let the complete data be $(Y_i, b_i, e_i), i = 1, \ldots, N$, since this gives an easy maximization for θ at the M-step. The observed data are just the Y_i's. It is then convenient to first obtain the joint distribution of (Y_i, b_i, e_i), as

$$\begin{pmatrix} Y_i \\ b_i \\ e_i \end{pmatrix} = N\left[\begin{pmatrix} X_i\beta \\ 0 \\ 0 \end{pmatrix}, \begin{pmatrix} Z_i D Z_i^T + \sigma^2 I & Z_i D & \sigma^2 I \\ D Z_i^T & D & 0 \\ \sigma^2 I & 0 & \sigma^2 I \end{pmatrix}\right].$$

Notice that $\text{cov}(Y_i, b_i, e_i)^T$ is singular, $(\text{cov}(Y_i, Z_i b_i + e_i) = I)$ and given (b_i, e_i), Y_i contributes nothing to the estimation of θ. However, the marginals, (Y_i, b_i) and (Y_i, e_i) are not singular and these are needed in our calculations.

As in ML estimation with the LMCD, given Σ, the ML estimate of β is WLS(Σ). Hence we derive likelihood equations for the θ parameters only. Thus the relevant piece of the complete data likelihood is

$$\prod_{i=1}^{N} \frac{e^{-b_i^T D^{-1} b_i/2}}{|D|^{1/2}} \quad \frac{e^{-e_i^T e_i/2\sigma^2}}{\sigma^{n_i}},$$

so that $\sum_{i=1}^N b_i b_i^T$ and $\sum_{i=1}^N e_i^T e_i$ are the complete data "sufficient statistics." It follows that the M-step for θ is

$$\widehat{D} = \sum_{i=1}^{N} b_i b_i^T \Big/ N$$

$$\widehat{\sigma}^2 = \sum_{i=1}^{N} e_i^T e_i \Big/ \Sigma n_i.$$

At the E-step, given $\widehat{\theta}$ and $\widehat{\beta} = \beta(\Sigma(\widehat{\theta})^{-1})$ calculate

$$S_1 = \sum_{i=1}^{N} E(b_i b_i^T | Y_i, \widehat{\theta}, \widehat{\beta})$$

and

$$S_2 = \sum_{i=1}^{N} E(e_i^T e_i | Y_i, \widehat{\theta}, \widehat{\beta}).$$

Given the joint moments of (Y_i, b_i, e_i), it is straightforward to show that the conditional moments of e_i and b_i given Y_i are:

(a) $E(b_i|Y_i) = 0 + DZ_i^T\Sigma_i^{-1}(Y_i - X_i\beta)$

(b) $\text{var}(b_i|Y_i) = D - DZ_i^T(\Sigma_i)^{-1}Z_iD$

(c) $E(b_ib_i^T|Y_i) = E(b_i|Y_i)E(b_i|Y_i)^{-1} + \text{var}(b_i|Y_i)$

$$= DZ_i^T\Sigma_i^{-1}(Y_i - X_i\beta)(Y_i - X_i\beta)^{-1}\Sigma_i^{-1}Z_iD + D - DZ_i^T\Sigma_i^{-1}Z_iD$$

(d) $E(e_i|Y_i) = 0 + \sigma^2\Sigma_i^{-1}(Y_i - X_i\beta)$

(e) $\text{var}(e_i|Y_i) = \sigma^2 I - \sigma^4\Sigma_i^{-1}$

(f) $E(e_ie_i^T|Y_i) = \sigma^4\Sigma_i^{-1}(Y_i - X_i\beta)(Y_i - X_i\beta)^T\Sigma_i^{-1} + \sigma^2\{I - \sigma^2\Sigma_i^{-1}\}$

(g) $E(e_i^Te_i|Y_i) = \text{tr } E\{e_ie_i^T|Y_i\}$.

It follows that the likelihood equations for $\widehat{\theta}_{ML}$ and $\widehat{\beta}_{ML}$ are the solution to the following:

$$\widehat{D} = \sum_{i=1}^{N} \left[\widehat{D}Z_i^T\widehat{\Sigma}_i^{-1}\left(Y_i - X_i\widehat{\beta}\right)\left(Y_i - X_i\widehat{\beta}\right)^T\widehat{\Sigma}_i^{-1}Z_i\widehat{D} + \widehat{D} - \widehat{D}Z_i^T\widehat{\Sigma}_i^{-1}Z_i\widehat{D} \right] \Big/ N, \tag{5.9}$$

$$\widehat{\sigma}^2 = \text{tr} \sum_{i=1}^{N} \left[\widehat{\sigma}^4\widehat{\Sigma}_i^{-1}\left(Y_i - X_i\widehat{\beta}\right)\left(Y_i - X_i\widehat{\beta}\right)^T\widehat{\Sigma}_i^{-1} + \sigma^2\left(I - \widehat{\sigma}^2\widehat{\Sigma}_i^{-1}\right) \right] \Big/ n_+ \tag{5.10}$$

and

$$\widehat{\beta} = \left(\Sigma X_i^T \widehat{\Sigma}_i^{-1} X_i\right)^{-1} \Sigma X_i^T \widehat{\Sigma}_i^{-1} Y_i \tag{5.11}$$

where

$$\widehat{\Sigma}_i = Z_i \widehat{D} Z_i^T + \widehat{\sigma}^2 I.$$

An iterative algorithm for $\theta^{(p)} \longrightarrow \theta^{(p+1)}$ can be obtained by evaluating the right-hand side of equations (5.9)–(5.10) at $(\beta^{(p)}, \theta^{(p)})$, and setting $\widehat{\sigma}^2$ and $\widehat{D}$ in the left-hand side to $\theta^{(p+1)}$ and using (5.11) to define $\beta^{(p+1)}$ given $\theta^{(p+1)}$. Using this EM algorithm to compute ML's is available as an option in some software packages but convergence can be very slow, especially when the between subject variance (D) is large relative to the within subject variance (σ^2). (This corresponds to a large fraction of missing information.) Alternatively, we can simply use "EM" approach to get expressions for first and second derivatives and then use Newton-Raphson for the computations. However, obtaining second derivatives for θ is more complex but needed for standard errors of $\widehat{\theta}$. See, for example, McCulloch and Searle (2001).

5.4 REML Estimation in the LMM

Recall that the "Bayesian" definition of REML is to maximize the marginal posterior likelihood giving β a flat prior:

$$\underbrace{\mathcal{L}(\theta; Y_1, \ldots, Y_N)}_{\text{REML likelihood}} = \int_{R^p} \underbrace{f(Y_1, \ldots, Y_N | \beta, \theta)}_{\text{ML likelihood}} \, d\beta.$$

This is just like an "incomplete data likelihood" where the complete data is $\mathcal{Z} = (Y_1, \ldots, Y_N, \beta)$ and the incomplete data is $\mathcal{Y} = (Y_1, \ldots, Y_N)$. From section 3.2, we have that

$$\frac{\partial \ln \mathcal{L}(\mathcal{Y}; \Phi)}{\partial \Phi} = E\left\{\frac{\partial \ln \mathcal{L}(\mathcal{Z}; \Phi)}{\partial \Phi} | \mathcal{Y}, \Phi\right\}$$

where $\mathcal{L}(\mathcal{Y}; \Phi)$ and $\mathcal{L}(\mathcal{Z}; \Phi)$ are the incomplete and complete data likelihoods. Thus for this case, we can write

$$\frac{\partial \ln \mathcal{L}(\theta; Y_1, \ldots, Y_N)}{\partial \theta} = E\left\{\frac{\partial \ln \mathcal{L}(\theta; Y_1, \ldots, Y_N, \beta)}{\partial \theta} | Y_1, \ldots, Y_N, \theta\right\},$$

where expectation is over the posterior distribution of β given the Y_i's and θ.

But $\partial \ln \mathcal{L}(\theta; Y_1, \ldots, Y_N, \beta)|\partial\theta$ gives the ordinary ML likelihood equations. Setting these likelihood equations to zero yields equations (5.9)–(5.10), for θ derived in the previous section, except that $\widehat{\beta}$ in (5.9–5.10) is just β, since the derivation is with respect to θ given β.

So to get REML equations, we take expectations of (5.9) and (5.10) (with all estimates replaced by their estimands) with respect to the posterior of β, given $Y_1, \ldots, Y_N$, which is distributed as

$$N\left(\widehat{\beta}, \left(\sum_{i=1}^{N} X_i^T \Sigma_i^{-1} X_i\right)^{-1}\right).$$

Because of joint normality, the conditional means of b_i and e_i are linear in β and the conditional variance does not depend upon β. Hence it is straightforward to see that taking expectations and setting $\theta = \widehat{\theta}_{\text{REML}}$ on both sides of (5.9), we obtain

$$\begin{aligned} \widehat{D}_{\text{REML}} = \sum_{i=1}^{N} \Big[& \widehat{D} Z_i^T \widehat{\Sigma}_i^{-1} (Y_i - X_i\widehat{\beta})(Y_i - X_i\widehat{\beta})^T \widehat{\Sigma}_i^{-1} Z_i \widehat{D} \\ & + \widehat{D} - \widehat{D} Z_i^T \widehat{\Sigma}_i^{-1} Z_i \widehat{D} \\ & + \widehat{D} Z_i^T \widehat{\Sigma}_i^{-1} X_i V X_i^T \widehat{\Sigma}_i^{-1} Z_i D \Big] \Big/ N, \end{aligned} \tag{5.12}$$

where $V = \text{var}(\beta|X_1, \ldots, Y_N)$ and $\widehat{\Sigma}_i$ and $\widehat{D}$ on the RHS are evaluated at $\widehat{D}_{\text{REML}}$ and $\widehat{\sigma}^2_{\text{REML}}$. Since the first three terms are identical to the expression for D_{ML}, and the last term is always positive, it is clear that the REML estimate is generally larger than the ML.

REMARK. We note a connection between the LMM and the LMCD. Suppose we are in the "true" missing data setting in the LMCD, where

$$Y_i = X_i\beta + e_i, \text{var}(e_i) = \Sigma_i,$$

and Σ is an unstructured $n \times n$ covariance matrix for person with complete data. How do these models relate? We can view the unstructured Σ as a special case of the LMM with $\sigma^2 = 0$ by taking $\underset{n_i \times n}{Z_i} = I_i$ to be the matrix of zeros and ones indicating missingness and $D_{n\times n} = \Sigma_{n\times n}$. Then

$$Y_i = X_i\beta + I_i b_i.$$

Here b_i is an $n \times 1$ vector of residuals for a person which might be viewed as "complete data residuals," i.e., e_i if all n observations are observed and

$$\begin{aligned} \text{var}(Y_i) &= Z_i D Z_i^T \\ &= I_i \Sigma I_i^T = \Sigma_i. \end{aligned}$$

Using this trick, we can use programs designed to fit variance component models to fit general (unstructured) Σ, but this is unnecessary as several programs are now available which fit unstructured Σ.

5.5 Estimating the Random Effects

Why estimate the random effects?

1. We might want individual growth curves. Examples include evaluating surrogate markers, and determining maximal growth rates for individuals (Donnelly *et al.*, 1995; Tsiatis *et al.*, 1996).

2. Random effects estimates can be used to establish value or cost. Examples include setting insurance premiums for small regions or small subgroups, establish breeding values for individual animals, establish optimal drilling locations in a region (Robinson, 1991; Cressie 1991).

3. Random effects estimates can be used to evaluate individual quality or performance. Examples include estimating both performance and outcomes for individual hospitals, event rates for local areas, etc.

In estimating individual effects it is sometimes advocated to treat the b_i's as fixed and use OLS to estimate an enlarged vector of fixed effects $\beta^* = (\beta, b_1, \ldots, b_N)$. This approach can be especially useful in the clustered data setting where b_i is a scalar and each n_i is substantial. However, this approach is of limited utility in a general setting for reasons discussed in Chapter 1, and because often many subjects will have small n_i and little information to estimate b_i. As we will discuss, the idea underlying random effects estimates is to "borrow" information from the whole in order to estimate individual effects.

Estimates of random effects are usually called *predictors* rather than estimates; they can be motivated in a Bayesian way (Empirical Bayes) or by extending the Gauss Markov theorem to include random effects. We take up the latter approach first. Recall that for the usual linear model

$$\underset{N\times 1}{Y} = \underset{N\times p}{X} \underset{p\times 1}{\beta} + \underset{N\times 1}{e}$$

where var $(e) = \sigma^2 I$, the **B**est **L**inear **U**nbiased **E**stimator (BLUE) of any estimable contrast $C^T\beta$ satisfies

$$E(C^T\widehat{\beta}) = C^T\beta$$

and

$$\mathrm{var}(C^T\widehat{\beta}) \le \mathrm{var}(C^T\beta^*)$$

for any other unbiased estimator β^*, where both $\widehat{\beta}, \beta^*$ are linear in Y, i.e., both estimates take the form AY for some A. To extend this to random effects, let B denote the vector of all random effects,

$$B^T = (b_1^T, \ldots, b_N^T)^T.$$

A BLUP is defined as a **B**est **L**inear **U**nbiased **P**redictor of

$$\lambda = C_1^T\beta + C_2^T B.$$

In this case both B and $Y = (Y_1^T, \ldots, Y_N^T)^T$ are treated as random variables. The BLUP has the same general properties, i.e., it is a linear function of the data Y, and it is unbiased and minimum variance in the class of unbiased linear predictors. It is not hard to show that for our case,

$$\widehat{b_i} = E(b_i|Y_i, \widehat{\beta}, \theta) = Z_i D \Sigma_i^{-1}(Y_i - X_i\widehat{\beta})$$

where $\widehat{\beta} = \widehat{\beta}(\Sigma(\theta)^{-1})$ is the BLUP estimator for fixed θ. Thus the BLUP estimator depends upon the unknown θ, which is typically estimated by ML or REML (Harville, 1977).

The empirical Bayes strategy is to regard each b_i as a random parameter with prior $N(0, D)$. Given data Y_i, we can compute the posterior of b_i as

$$b_i|Y_i \sim N\left(\mu_{b_i}, \Sigma_{b_i}\right)$$

where

$$\mu_{b_i} = E\left(b_i|Y_i, \beta, \theta\right) = DZ_i^T\Sigma_i^{-1}\left(Y_i - X_i\beta\right)$$

and

$$\Sigma_{b_i} = \mathrm{var}\left(b_i|Y_i, \beta, \theta\right) = D - DZ_i^T\Sigma_i^{-1}Z_iD.$$

So (if we knew β, θ) an optimal "Bayes" estimate of b_i would be μ_{b_i} (both the posterior mean and the mode). Unlike most Bayes settings, we have replication (in the form of Y_i), which enables us to estimate both θ and β from the data, so the "empirical" Bayes estimate of b_i takes the form

$$\widehat{b_i} = \mu_{b_i}|_{\widehat{\beta},\widehat{\theta}} = \widehat{D}Z_i^T\widehat{\Sigma}_i^{-1}(Y_i - X_i\widehat{\beta}).$$

This is identical to the BLUP estimator when θ is estimated in the same way.

Obtaining a valid estimate for var (b_i) is a difficult problem. We do not want to use

$$\mathrm{var}\left(b_i|Y_i, \widehat{\beta}, \widehat{\theta}\right)$$

because it underestimates the variability arising due to uncertainty about $\widehat{\beta}$. Uncertainty about θ usually has a second order effect (at least for large samples), so it is uncertainty about $\widehat{\beta}$ that is more problematic. The frequentist approach is to use

$$\operatorname{var}\left(\widehat{b}_i - b_i|\widehat{\beta}, \widehat{\theta}\right)$$

since both $\widehat{b_i}$ and b_i are random variables. The Bayesian approach is to use the marginal posterior of b_i, integrating out β. These two again yield identical variance expressions for the linear model:

$$\begin{aligned}\operatorname{var}\left(\widehat{b}_i - b_i|Y_i, \beta, \theta\right) &= \operatorname{var}(b_i|Y, \theta) \\ &= D - DZ_i^T\Sigma_i^{-1}Z_iD \\ &\quad + DZ_i^T\Sigma_i^{-1}X_i\left(\Sigma X_i^T\Sigma_i^{-1}X_i\right)^{-1}X_i^T\Sigma_i^{-1}Z_iD.\end{aligned}$$

This expression often used to get standard errors for EB estimates even though it does ignore error of estimation in $\widehat{\theta}$. Note that only estimates for σ^2 and D are needed to evaluate the variance expression, since β has been integrated out.

Finally, notice that the EB estimates are sometimes called shrinkage (or James–Stein) estimators. To see why, return to the two-stage model where

$$Y_i = A_iZ_i\beta + Z_ib_i + e_i$$

and

$$\beta_i = A_i\beta + b_i,$$

and assume Z_i is of full rank. After much algebra, we can show that

$$\widehat{\beta}_i^{\text{EB}} = A_i\widehat{\beta} + \widehat{b}_i = W_i\widehat{\beta}_i^{\text{OLS}} + (I - W_i)A_i\widehat{\beta},$$

where

$$\widehat{\beta} = \widehat{\beta}\left(\Sigma(\theta)^{-1}\right), \quad \widehat{\beta}_i^{\text{OLS}} = \left(Z_i^TZ_i\right)^{-1}Z_i^TY_i,$$

and W_i is the variance ratio

$$W_i = D\left(D + \sigma^2(Z_i^TZ_i)^{-1}\right)^{-1}.$$

If $\sigma^2 = 0$, meaning that we have perfect information about β_i from Y_i, we see that $\widehat{\beta}_i^{\text{EB}} = \widehat{\beta}_i^{\text{OLS}}$. Alternatively, if $D = 0$, there is no variability in the individual random effects, and $\widehat{\beta}_i^{\text{EB}} = A_i\widehat{\beta}$.

This approach can also be used to get "smoothed" or predicted values as

$$\widehat{Y}_i = X_i\widehat{\beta} + Z_i\widehat{b}_i,$$

for the observed Y_i's as well as for values of Y_{ij} that are missing. For example, if we fit straight lines (b_i is intercept and slope), then $(\widehat{\beta}, \widehat{b}_i)$ can be used to get each individual's curve and predicted values. If the mean response is estimated only at the points specified by the protocol, the empirical Bayes approach can be used to obtain estimates of any missing Y_{ij}'s.

Chapter 6

Longitudinal Data Analysis for Counts and Binary Outcomes: Generalized Estimating Equations (GEE)

In many settings, the outcomes recorded on individuals are counts or binary responses. In this chapter we extend the theory in the preceding chapters to permit a regression analysis which does not require the mean responses to be linear in X_i. In the univariate setting, the generalized linear model (GLM, McCullagh and Nelder, 1989) offers an approach which unifies linear, log and logistic regression analysis. It was extended to the distribution free multivariate setting by Liang and Zeger (1988) and Zeger and Liang (1988). Extensions of the likelihood approach and the random effects models to the nonlinear setting is more complex; we will review some of the suggested approaches in subsequent chapters. In this chapter, we begin by reviewing the basic ideas for GLM's in the univariate setting, and then discuss the GEE extension to correlated data.

6.1 The Generalized Linear Model (GLM) for Univariate Outcomes.

Suppose now that Y_i is a scalar outcome, X_i is a $1 \times p$ row vector of covariates, β is a $p \times 1$ vector of regression coefficients and

$$\mu_i = E(Y_i) = g(X_i\beta), \tag{6.1}$$

where

$$g^{-1}(\mu_i) \equiv \ell(\mu_i) = X_i\beta.$$

Here $g(\cdot)$ and $\ell(\cdot)$ are known functions; $\ell(\cdot)$ is called the link function and $g(\cdot)$ is the inverse link function.

Examples. For the linear model, both $\ell(\cdot)$ and $g(\cdot)$ are the identity functions: $\ell(\mu_i) = \mu_i$, and $\ell(\cdot)$ is called the identity link. If Y_i is a count, so that $\mu_i > 0$, a natural link function is the log:

$$\log\ \mu_i = X_i\beta \Rightarrow \mu_i = e^{X_i\beta}.$$

Here $\ell(\cdot)$ is the log link. With binary data, $E(Y_i) = P(Y_i = 1)$, hence $0 < \mu_i < 1$ and a popular link function is

$$\text{logit}\mu_i = \log\left(\mu_i/(1-\mu_i)\right) = X_i\beta,$$

or

$$\mu_i = e^{X_i\beta}/(1 + e^{X_i\beta}).$$

Here $\ell(\cdot)$ is the logit link.

As in the LMCD setting, it is possible to implement a distribution free analysis using only the assumption of the mean model (6.1), or we may fully specify the distribution of Y_i (possibly as a function of other parameters) and use a fully parametric analysis. The distribution free approach estimates β by minimizing the objective function

$$Q(\beta) = \sum_{i=1}^{N} W_i(Y_i - \mu_i)^2$$

for some arbitrary choice of weights, W_i. Straightforward differentiation of $Q(\beta)$ with respect to β gives a $p \times 1$ vector of derivatives $\partial Q(\beta)/\partial\beta_j, j = 1, \ldots, p$:

$$\frac{\partial Q(\beta)}{\partial\beta} = 2\sum_{i=1}^{N}\left(\frac{\partial\mu_i}{\partial\beta}\right) W_i\left(Y_i - \mu_i\right); \tag{6.2}$$

setting (6.2) equal to zero gives $\widehat{\beta}$, since given β,

$$\widehat{\mu}_i = g(X_i\widehat{\beta}),$$

and the weights are assumed known. This can be viewed as a semiparametric approach because:

i) Any estimator of β that is consistent and asymptotically normal, assuming *only* (6.1) is true, is asymptotically equivalent to $\widehat{\beta}(W)$ for some choice of W.

ii) The choice of weights which gives $\widehat{\beta}(W)$ the smallest variance among estimators in this class is $W_i^{-1} = \text{var}(Y_i|X_i) = V_i$.

iii) The asymptotic distribution of $\widehat{\beta}(W)$ satisfies

$$\sqrt{N}\left(\widehat{\beta}(W) - \beta\right) \to N(0, C)$$

where

$$C = \lim_{N\to\infty} I_0^{-1} I_1 I_0^{-1}, \tag{6.3}$$

$$I_0 = \left[\sum_{i=1}^{N} \left(\frac{\partial \mu_i}{\partial \beta}\right) W_i \left(\frac{\partial \mu_i}{\partial \beta}\right)^T\right] \Big/ N,$$

and

$$I_1 = \left[\sum_{i=1}^{N} \left(\frac{\partial \mu_i}{\partial \beta}\right) W_i V_i W_i \left(\frac{\partial \mu_i}{\partial \beta}\right)^T\right] \Big/ N.$$

A consistent estimator of C is obtained by evaluating $(\partial\mu_i/\partial\beta)$ at $\widehat{\beta}$, and substituting $(Y_i - \widehat{\mu}_i)^2$ for V_i. The same asymptotic limiting distribution will obtain when the W_i are replaced by estimated $\widehat{W}_i$.

In the GLM, we additionally assume that

$$V_i = \text{var}(Y_i|X_i) = V(\mu_i)\phi,$$

where $V(\mu_i)$ is a known function depending upon the mean and ϕ is a known or unknown scalar factor. This implies that V_i depends upon the covariates X_i only through the mean μ_i.

Examples. With Y_i binary, $\text{var}(Y_i) = \mu_i(1 - \mu_i) = V(\mu_i)$ and $\phi = 1$. If we assume a Poisson variance for count data Y_i, var $(Y_i) = \mu_i\phi$, where

ϕ is a dispersion parameter. In the linear case, we usually assume the variance does not depend upon μ_i, and take $V(\mu_i) = 1$ and $\sigma^2 = \phi$.

Notice that if

$$W_i = (V(\mu_i)\phi)^{-1},$$

ϕ drops out of the estimating equations, so we may equivalently take

$$W_i = (V(\mu_i))^{-1},$$

so that the estimating equations become

$$\sum_{i=1}^{N} \left(\frac{\partial \mu_i}{\partial \beta}\right)_{\big|\widehat{\beta}} V(\widehat{\mu}_i)^{-1}(Y_i - \widehat{\mu}_i) = 0. \tag{6.4}$$

Under this assumption, the same limiting distribution holds, with now

$$\text{var}\sqrt{N}(\widehat{\beta} - \beta) = C$$

for

$$C = \lim_{N\to\infty} \left[\frac{1}{N}\sum_{i=1}^{N} \left(\frac{\partial \mu_i}{\partial \beta}\right) V_i^{-1} \left(\frac{\partial \mu_i}{\partial \beta}\right)^T\right]^{-1}, \tag{6.5}$$

and

$$\widehat{C} = \left[\frac{1}{N}\sum_{i=1}^{N} \left(\frac{\partial \mu_i}{\partial \beta}\right)_{\big|\widehat{\beta}} \widehat{V}_i^{-1} \left(\frac{\partial \mu_i}{\partial \beta}\right)^T_{\big|\widehat{\beta}}\right]^{-1}.$$

REMARKS.

(i) The GLM assumes that the model $\text{var}(Y_i) = V(\mu_i)\phi$ is correct for $\text{var}(Y_i) = V_i$; once we estimate β, we have an estimate for V_i (up to a proportionality constant).

(ii) Equation (6.5) for the $\widehat{\text{var}}(\sqrt{N}(\widehat{\beta} - \beta))$ is only consistent when $\text{var}(Y_i) = V(\mu_i)\phi$ is correct.

(iii) The estimating equations in (6.4) are called quasi-likelihood score equations (Weddeburn, 1974; McCullagh and Nelder, 1989).

(iv) Suppose further that Y_i has an exponential family distribution with canonical parameter θ_i, so that

$$f(Y_i) = e^{(Y_i\theta_i + \tilde{a}(\theta_i) + C(Y_i))/\phi},$$

where μ_i is some function of θ_i, and

$$\tilde{a}(\theta_i)/\phi = -\ln \int e^{(t\theta_i + C(t))/\phi} dt.$$

For this family of distributions, it is easily shown that

$$E(Y_i) = -\partial\tilde{a}(\theta_i)/\partial\theta_i$$

and

$$\text{var}\,(Y_i) = -\partial^2\tilde{a}(\theta_i)/\partial^2\theta_i.$$

We now show that the quasi-likelihood equations correspond exactly to the likelihood score equations. Here we treat ϕ as a fixed scale parameter. Note that (iv) implies

$$\mathcal{L}_\phi\,(\beta)) = \prod_{i=1}^{N} f\,(Y_i) \propto e^{\left(\sum_{i=1}^{N} Y_i\theta_i + \sum_{i=1}^{N}\tilde{a}(\theta_i)\right)/\phi}$$

so that

$$\frac{\partial \ln \mathcal{L}_\phi(\beta)}{\partial\beta} = \left[\sum_{i=1}^{N} Y_i\left(\frac{\partial\theta_i}{\partial\beta}\right) + \sum_{i=1}^{N}\left(\frac{\partial\tilde{a}(\theta_i)}{\partial\beta}\right)\right]\frac{1}{\phi}.$$

Using the chain rule we have that

$$\frac{\partial\tilde{a}(\theta_i)}{\partial\beta} = \frac{\partial\tilde{a}(\theta_i)}{\partial\theta_i}\,\frac{\partial\theta_i}{\partial\beta} = -\mu_i\frac{\partial\theta_i}{\partial\beta}.$$

But

$$\frac{\partial\mu_i}{\partial\beta} = \frac{\partial\mu_i}{\partial\theta_i}\,\frac{\partial\theta_i}{\partial\beta} = -\frac{\partial^2\tilde{a}(\theta_i)}{\partial\theta_i^2}\left(\frac{\partial\theta_i}{\partial\beta}\right) = \text{var}(Y_i)\left(\frac{\partial\theta_i}{\partial\beta}\right),$$

so that

$$\frac{\partial\theta_i}{\partial\beta} = \text{var}(Y_i)^{-1}\frac{\partial\mu_i}{\partial\beta}$$

and

$$\frac{\partial\tilde{a}(\theta_i)}{\partial\beta} = -\mu_i\text{var}(Y_i)^{-1}\frac{\partial\mu_i}{\partial\beta}.$$

Now using the fact $\text{var}(Y_i) = V(\mu_i)\phi$, we have that

$$\frac{\partial \ln \mathcal{L}_\phi(\beta)}{\partial\beta} \propto \sum_{i=1}^{N}\left(\frac{\partial\mu_i}{\partial\beta}\right)V(\mu_i)^{-1}(Y_i - \mu_i). \tag{6.6}$$

This shows that the likelihood equations are equal to the quasi-likelihood score equations when Y_i has an exponential family density with specified mean and variance, and $W_i = V(\mu_i)^{-1}$.

With exponential families, θ_i is the canonical parameter. We can use it to define the canonical link. If

$$\ell\,(\mu_i) = \theta_i = X_i\beta$$

then ℓ is said to be the canonical link. In this case, the likelihood can be further simplified by noting that

$$\frac{\partial \mu_i}{\partial \beta} = \frac{\partial \mu_i}{\partial \theta_i}\frac{\partial \theta_i}{\partial \beta} = \frac{\partial \mu_i}{\partial \theta_i} X_i^T$$

and

$$\frac{\partial \mu_i}{\partial \theta_i} = -\frac{\partial^2 \tilde{a}(\theta_i)}{\partial \theta_i^2} = \text{var}(Y_i)$$

hence the likelihood equations become simply

$$\sum_{i=1}^{N} X_i^T \left(Y_i - \widehat{\mu}_i\right) = 0.$$

To continue with the likelihood approach assuming the exponential family density (iv), the asymptotic variance of $\widehat{\beta}$ is given by the expected value of $-\partial^2 \ln \mathcal{L}_\phi(\beta)/\partial\beta\partial\beta^T$. Differentiating (6.6) with respect to β^T, we see that only one term has nonzero expectation:

$$-E\left(\frac{\partial \ln \mathcal{L}_\phi(\beta)}{\partial\beta\partial\beta^T}\right) = \sum_{i=1}^{N} \frac{\partial \mu_i}{\partial \beta} \left(\text{var}(Y_i)\right)^{-1} \left(\frac{\partial \mu_i}{\partial \beta}\right)^T,$$

hence

$$A \text{ var } \widehat{\beta} = \left\{\sum_{i=1}^{N} \left(\frac{\partial \mu_i}{\partial \beta}\right) \left(\text{var}(Y_i)\right)^{-1} \left(\frac{\partial \mu_i}{\partial \beta}\right)^T\right\}^{-1}$$

in agreement with (6.5). As we will see, the estimating equation and likelihood based approaches generally do not coincide in the multivariate setting with generalized linear models.

6.2 Generalized Linear Models for Longitudinal Data

As before, we will assume an $n_i \times 1$ vector of outcomes, Y_i, where any missingness in the data are MCAR. In addition, each observation is assumed to have a $p \times 1$ vector of covariates X_{ij} so that

$$E\left(Y_{ij}\right) = \mu_{ij} = g\left(X_{ij}^T\beta\right)$$

and

$$\ell\left(\mu_{ij}\right) = X_{ij}^T\beta$$

for some suitable link function $\ell(\cdot)$. Thus we may write

$$E(Y_i) = \mu_i$$

where

$$\ell(\mu_i) = X_i\beta$$

and $\ell(\mu_i)$ denotes the vector: $(\ell(\mu_{i1}), \ldots, \ell(\mu_{in_i}))^T$. This model is the natural extension of the longitudinal data model considered in the linear setting, with the only difference being that we allow for a generalized link-function linking the mean response vector μ_i to the covariates. In all other respects, the two are similar; it permits unbalanced designs, unequal clusters, etc. It is sometimes referred to as a marginal model to emphasize the point that the means, μ_{ij}, are marginal for each Y_i:

$$E(Y_{ij}|X_i\beta) = \mu_{ij}.$$

In this respect it does not differ from the linear model case.

Example 1. The Harvard Six-Cities Study of Air Pollution and Health gathered data annually on school children in six cities. One outcome studied was the presence or absence of respiratory illness in the preceding year. The relationship between maternal smoking status and rates of respiratory illness was one feature of interest in the study. Here each child has four annual indicators

$$\begin{aligned} Y_{ij} &= 1 \quad \text{if illness is past year} \\ &= 0 \quad \text{otherwise} \end{aligned}$$

and we assume

$$\text{logit}(\mu_{ij}) = \beta_0 + \beta_1 MS_i + \beta_2 \text{ age}_{ij} + \beta_3 MS_i \times \text{ age}_{ij}$$

where $MS_i = 1$ if mother smoked at the beginning of the study, 0 otherwise and age_{ij} is the age of the ith child at the jth occasion.

Example 2. Diggle *et al.* (1994) describe a clinical trial of progabide in the treatment of epileptic seizures. Patients were randomized to progabide (31 patients) or placebo (28 patients), and measured at baseline, and every two weeks until week 8. Responses were number of seizures in each period. Covariates include baseline seizure rate, period and treatment group. Here Y_{ij} is a count of the number of seizures for the ith subject in the jth period, $j = 1, \ldots, 4$. We assume

$$\log \mu_{ij} = X_{ij}^T\beta$$

where X_{ij} can include baseline seizure counts (perhaps transformed), treatment, period and treatment $\times$ period.

It is natural in this setting to further assume that

$$\text{var}(Y_{ij}) = V(\mu_i)\phi$$

for suitable $V(\cdot)$ because with count and binary data, the variance does typically depend upon the mean. For example, if Y_{ij} is binary, then by definition $V(\mu_{ij}) = \mu_{ij}(1 - \mu_{ij})$ and $\phi = 1$. With count data, $\text{var}(Y_{ij}) = \mu_i$ can be a rather strong assumption derived from Poisson theory. Over dispersion, $\phi > 1$, implies $\text{var}(Y_{ij}) > \mu_i$, so this can be a more reasonable model. In the multivariate setting

$$\text{var}(Y_i) = \phi A_i^{1/2} R_i(\alpha) A_i^{1/2}$$

where $A_i = \text{diag}\{V(\mu_{ij})\}$. Now however, there is no natural set of assumptions as to how $R_i(\alpha)$ should depend upon μ_i. Thus we will leave $R_i(\alpha)$ as unspecified. As before, we let the true variance of Y_i be denoted by Σ_i, and let

$$V_i \propto W_i^{-1} = A_i^{1/2} R_i(\alpha) A_i^{1/2}$$

denote a "working variance" assumption. Some authors refer to $R_i(\alpha)$ as a "working correlation matrix," implicitly assuming the variance assumption is correct, but not necessarily $R_i(\alpha)$.

6.3 Estimation via GEE.

The basic GEE strategy is to simply generalize the quasi-likelihood equations to the multivariate setting by replacing Y_i and μ_i by their vector counterparts, and using a weight matrix W_i. This yields

$$\sum_{i=1}^{N} \left(\frac{\partial \mu_i}{\partial \beta} \right)^T_{\Big|\widehat{\beta}} \widehat{W}_i \left(Y_i - \widehat{\mu}_i \right) = 0.$$

Here we define $\partial\mu_i/\partial\beta$ as an $n_i \times p$ matrix whose jth row is $\partial\mu_{ij}/\partial\beta^T$. Although optimally we would take $W_i = V_i^{-1}$, in fact any positive definite and symmetric matrix can by used for W_i. If $W_i = V_i^{-1}$, it now depends upon β, but $R_i(\alpha)$ can be specified arbitrarily provided that each W_i remains symmetric and positive definite. In fact, if $R_i(\alpha) = I$, then the GEE just reduces to the GLM analysis treating all Y_{ij} as independent observations, e.g., if $V(\mu_{ij}) = \mu_{ij}(1 - \mu_{ij})$ and $\ell(\cdot)$ is the logit (or probit) link, then GEE reduces to ordinary logistic (or probit) regression, treating all Y_{ij} as independent. Notice also, that for the identity link and $W_i = \Sigma_i^{-1}$, the GEE reduces to the multivariate normal likelihood equations.

We have the following property for $\widehat{\beta}$: $\sqrt{N}(\widehat{\beta}(W) - \beta) \longrightarrow N(0, C)$, where

$$C = \lim_{N\to\infty} I_0^{-1} I_1 I_0^{-1},$$

$$I_0 = \lim_{N\to\infty} \left[\sum_{i=1}^{N} \left(\frac{\partial \mu_i}{\partial \beta}\right)^T W_i \left(\frac{\partial \mu_i}{\partial \beta}\right) \right] \Big/ N, \tag{6.7}$$

$$I_1 = \lim_{N-\infty} \left[\sum_{i=1}^{N} \left(\frac{\partial \mu_i}{\partial \beta}\right)^T W_i \Sigma_i W_i \left(\frac{\partial \mu_i}{\partial \beta}\right) \right] \Big/ N.$$

As before, asymptotic efficiency of estimation is best with $W_i = \Sigma_i^{-1}$.

The GEE equations can be simplified as follows. First define

$$\frac{\partial \mu_i}{\partial \beta} = \begin{pmatrix} \frac{\partial \mu_{i1}}{\partial \beta_1} & \cdots & \frac{\partial \mu_{i1}}{\partial \beta_p} \\ \vdots & & \vdots \\ \frac{\partial \mu_{in_i}}{\partial \beta_1} & \cdots & \frac{\partial \mu_{in_i}}{\partial \beta_p} \end{pmatrix}_{n_i \times p} .$$

Recall that $\ell(\mu_{ij}) = X_{ij}^T \beta = \ell_{ij}$, thus

$$\frac{\partial \mu_{ij}}{\partial \beta} = \frac{\partial \mu_{ij}}{\partial \ell_{ij}} \frac{\partial \ell_{ij}}{\partial \beta} = \frac{\partial \mu_{ij}}{\partial \ell_{ij}} X_{ij},$$

so that

$$\left(\frac{\partial \mu_i}{\partial \beta}\right)^T = X_i^T \Delta_i \text{ with } \Delta_i = \text{ diag } \left\{\frac{\partial \mu_{ij}}{\partial \ell_{ij}}\right\},$$

and the GEE equations become

$$\sum_{i=1}^{N} X_i^T \widehat{\Delta}_i \widehat{W}_i (Y_i - \mu_i) = 0.$$

Examples. If ℓ_{ij} is the identity link, then $\ell_{ij} = \mu_{ij}$ and $\Delta_i = I$ and we have generalized least squares. If $\ell_{ij} = \log(\mu_{ij}/(1 - \mu_{ij}))$, then $\partial \mu_{ij}/\partial \ell_{ij} = \mu_{ij}(1 - \mu_{ij}) = V(\mu_{ij})$, and $\Delta_i = \text{ diag } (\mu_{ij}(1 - \mu_{ij}))$.

If, in addition, we assume that marginally, each Y_{ij} follows the exponential family density, with canonical parameter θ_i then

$$\begin{aligned} \frac{\partial \mu_{ij}}{\partial \ell_{ij}} &= \frac{\partial \mu_{ij}}{\partial \theta_{ij}} \frac{\partial \theta_{ij}}{\partial \ell_{ij}} \\ &= b_{ij} \operatorname{var}(Y_{ij}) \qquad \text{for } b_{ij} = \frac{\partial \theta_{ij}}{\partial \ell_{ij}}, \end{aligned}$$

and we can write $\Delta_i = B_i A_i$, where $B_i =$ diag $\{b_{ij}\}$, and A_i is diag$\{Y_i\}$, so that

$$\left(\frac{\partial \mu_i}{\partial \beta}\right)^T = \underset{p\times n_i}{X_i^T} \quad \underset{n_i\times n_i}{B_i} \quad \underset{n_i\times n_i}{A_i} .$$

Note that if the canonical link is used $\ell_{ij} = \theta_{ij}$ and $B_i = I$, and further, if $R(\alpha) = I$, so that

$$\sum_{i=1}^{N} X_i^T (Y_i - \mu_i) = 0.$$

6.4 Estimating the Correlation Matrix.

Assuming that var$(Y_{ij}) = V(\mu_{ij})\phi$ where V is known, the parameters in A_i will be determined by $\widehat{\beta}$, thus to estimate W_i, it remains only to model and estimate α. Models for the correlation are not different from those considered in the linear setting (except for random effects models to be considered later), i.e., we may choose unstructured (in the balanced setting), compound symmetry, serial correlation models, etc. Zeger and Liang (1984) proposed the following procedure to estimate α:

i) Estimate β by setting $R_i(\alpha) = I$ to get $\widehat{\beta}_I$ (independence working assumption).

ii) Obtain an estimate of α using the normalized residuals $A_i^{1/2}(Y_i - \widehat{\mu}_i)$, with $\widehat{\mu}_i$ evaluated at $\widehat{\beta}_I$. The details of this step depend upon the model for α and degree of balance in the data. Call this $\widehat{\alpha}_1$.

iii) Set $\widehat{W}_i^{-1} = \widehat{A}_i^{1/2} R_i(\widehat{\alpha}_1) \widehat{A}_i^{1/2}$ and use GEE to get $\widehat{\beta}_1$, holding $R_i(\widehat{\alpha}_1)$ fixed. Here $\widehat{A}_i$ depends upon $\widehat{\mu}_i$.

Iterate ii) and iii) to convergence. In practice, one step is often used, and may, in fact, be preferable if estimates of $R_i(\alpha)$ are unstable due to sparse data or small sample sizes. To compute $\widehat{\beta}$ given a fixed α we can use Fisher Scoring.

Estimating the α parameter can be done using the same method-of-moment approach used in the semi-parametric linear model setting. First consider the balanced and complete case with $n_i = n$ and unstructured $R(\alpha)$. As before, $A_i =$ diag $V(\mu_i)$, so that A_i depends only on β. Given $\widetilde{\beta}$, we may estimate α and ϕ as follows. Let $\widetilde{\mu}_i$ denote μ_i evaluated at $\widetilde{\beta}$, and

$$\widetilde{V}_i = (Y_i - \widetilde{\mu}_i)(Y_i - \widetilde{\mu}_i)^T,$$

so in large samples we have (assuming $\text{var}(Y_i) \doteq \phi A_i^{1/2} R(\alpha) A_i^{1/2}$):

$$E\left(\widetilde{V}_i\right) \doteq \phi \widetilde{A}_i^{1/2} R(\alpha) \widetilde{A}_i^{1/2}.$$

Thus we take

$$\begin{aligned} \widehat{R}(\widehat{\alpha}) &= \sum_{i=1}^{N} \left\{ \widetilde{A}_i^{-1/2} \widetilde{V}_i \widetilde{A}_i^{-1/2} \right\} \Big/ \widehat{\phi} N^* \\ \text{and} \qquad & \qquad\qquad\qquad\qquad\qquad\qquad\qquad\qquad (6.8) \\ \widehat{\phi} &= \sum_{i=1}^{N} \frac{\widetilde{\psi}_i^T \widetilde{\psi}_i}{N^*} \qquad \text{for } \widetilde{\psi}_{ij} = (Y_{ij} - \widetilde{\mu}_{ij}) / A_{ij}^{1/2}, \end{aligned}$$

where $N^* = \sum_{i=1}^{N} n_i$. Notice that ϕ is a constant variance inflation function for $V(\mu_{ij})$. In practice, using (6.8) yields an $\widehat{R}$ such that $\text{diag}(\widehat{R})$ will not be all ones, so they are usually forced to one at each iteration. This is because the variance terms are estimated from the model, but the covariance parameters are not.

We can formalize this method-of-moments estimation of α by using a similar set of estimating equations for an arbitrary $R_i(\alpha)$. This is convenient for the setting where the occasions of measurement may vary from subject to subject, but the correlations can be modeled with a limited number of parameters. Let

$$\rho_{ijk} = \rho_{ijk}(\alpha)$$

where

$$\rho_{ijk} = \text{corr}\,(Y_{ij}, Y_{ik}) \ ,$$

and

$$r_{ijk} = \phi (Y_{ij} - \mu_{ij})(Y_{ik} - \mu_{ik}) / A_{ij}^{1/2} A_{ik}^{1/2}$$

so that

$$E(r_{ijk}) = \rho_{ijk} \ .$$

Then estimating equations for α are given by

$$\sum_{i=1}^{N} C_i^T U_i^{-1} (r_i - \rho_i) = 0$$

where

$$\begin{aligned} \rho_i^T &= \left(\rho_{i12}, \ldots, \rho_{in_i(n_i-1)}\right)^T, \\ r_i^T &= \left(r_{i12}, \ldots, r_{in_i(n_i-1)}\right), \\ C_i &= \frac{\partial \rho_i}{\partial \alpha} \qquad \text{and} \qquad U_i = \text{var}(r_i) \ . \end{aligned}$$

Note that the dimension of r_i and ρ_i is $n_i(n_i-1)/2 = m_i$. Specifying U_i for optimal estimation requires both the third and fourth moments of Y_{ij}, so usually we set $U_i = I_{m_i}$, to give

$$\Sigma C_i^T (r_i - \rho_i) = 0 \; .$$

When the two sets of estimating equations are used to estimate β and α, solving them iteratively but separately we have:

Given (α^k, β^k):

i) Fix α^k, solve GEE equations to get β^{k+1}, where

$$W_i^k = \left(A_i^{1/2} R_i(\alpha^k) A_i^{1/2}\right)^{-1} \; .$$

ii) Fix β^{k+1}, solve for $\alpha^{(k+1)}$ using the α estimating equations, where $\mu_i, V(\mu_i)$ are evaluated at β^{k+1}; ϕ is estimated as before.

Comments.

1. In many cases, especially with $U_i = I$, the α estimating equations can be solved non-iteratively.
2. When the data are highly unbalanced, these estimating equations for α may not be so attractive.
3. One difficulty encountered with binary data is that the correlations are not a natural measure of association as they are in the linear model setting. In particular, the correlation is restricted by the range of the data; all values between -1 and $+1$ are generally not possible.

To elaborate on point 3, consider two binary variables Y_1, Y_2, with means μ_1, μ_2. Then

$$\text{corr} \;=\; \frac{E(Y_1 Y_2) - \mu_1\mu_2}{\sqrt{\mu_1(1-\mu_1)\mu_2(1-\mu_2)}}$$

. But $E(Y_1Y_2) = \Pr(Y_1 = Y_2 = 1) = \mu_{11}$ where

<table>
<tr><td></td><td></td><td colspan="2">Y_2</td><td></td></tr>
<tr><td></td><td></td><td>1</td><td>0</td><td></td></tr>
<tr><td rowspan="2">Y_1</td><td>1</td><td>μ_{11}</td><td>$\mu_1 - \mu_{11}$</td><td>μ_1</td></tr>
<tr><td>0</td><td>$\mu_2 - \mu_{11}$</td><td>$1 - \mu_2 - \mu_1 + \mu_{11}$</td><td>$(1-\mu_1)$</td></tr>
<tr><td></td><td></td><td>μ_2</td><td>$(1-\mu_2)$</td><td>1</td></tr>
</table>

The maximum value of μ_{11} is $\min(\mu_1, \mu_2)$. Assume $\mu_1 < \mu_2$, then the max of $\mu_{11} = \mu_1$ and

$$\begin{aligned}\text{corr} &= \frac{\mu_1 - \mu_1\mu_2}{\sqrt{\mu_1(1-\mu_1)\mu_2(1-\mu_2)}} = \frac{\mu_1(1-\mu_2)}{\sqrt{\mu_1(1-\mu_1)\mu_2(1-\mu_2)}} \\ &= \frac{\sqrt{\mu_1(1-\mu_2)}}{\sqrt{(1-\mu_1)\mu_2}} < 1\end{aligned}$$

because $\mu_1 < \mu_2 \Rightarrow (1-\mu_2) < (1-\mu_1)$. The correlation can attain one only if $\mu_1 = \mu_2$.

With binary response, and sometimes with count data as well, we often use odds ratios to describe association:

$$\text{OR} = \frac{P(Y_1 = Y_2 = 1)(P(Y_1 = Y_2 = 0)}{P(Y_1 = 1, Y_2 = 0)P(Y_1 = 0, Y_2 = 1)}.$$

This is an desirable measure of association for a variety of reasons:

1. OR $= 1$ or ln OR $= 0$ implies (Y_1, Y_2) are independent.
2. ln (OR) is symmetric about 0 and unbounded; it is not constrained by the marginal moments of (Y_1, Y_2).
3. It is invariant to marginal specification of μ_1 and μ_2. That is, any (μ_1, μ_2) pair is compatible with any value of OR; this explains its appeal in case-control studies.

Various authors (Prentice, 1988; Lipsitz *et al.*, 1990, 1991; Liang *et al.*, 1992) have suggested replacing the α estimating equations by a set of odds-ratio estimating equations. The idea is that in the 2×2 table, if the margins are fixed (μ_1, μ_2), there is one remaining degree-of-freedom for determining association. We can use it to estimate the odds ratio, then calculate the correlation needed for the β equations as a function of the odds ratio. This approach has some attractive features, but has limitations when there are more than two responses $(n_i > 2)$. In this case, the set of $n_i(n_i - 1)/2$ odds ratios are given by

$$\Omega_{ijk} = \frac{P(Y_{ij} = 1,\ Y_{ik} = 1)\ P(Y_{ij} = 0, Y_{ik} = 0)}{P(Y_{ij} = 1, Y_{ik} = 0)\ P(Y_{ij} = 0, Y_{ik} = 1)}.$$

In particular, the parameter space of the Ω_{ijk}'s, like that of the ρ_{ijk}'s depends upon the μ_i.

Liang *et al.* (1992) proposed an extension to GEE termed GEE-1 which can be used to estimate β and α for an arbitrary parameterization of the association, say η_{ijk}, that permits a unique transformation from η_i to ρ_i. This permits one to obtain an expression for W_i in terms of $R_i(\alpha)$. Let $S_{ijk} = (Y_{ij} - \mu_{ij})(Y_{ik} - \mu_{ik})$ and $E(S_{ijk}) = \eta_{ijk} = E(Y_{ij}Y_{ik}) - \mu_{ij}\mu_{ik}$. Then the association parameters indexed by α can be defined in terms of η_{ijk} and μ_i.

For example,

$$\rho_{ijk} = \eta_{ijk} / \left(\mu_{ij}(1-\mu_{ij})\mu_{ik}(1-\mu_{ik})\right)^{1/2}$$

and the odds ratio Ω_{ijk} can be expressed as:

$$\Omega_{ijk} = \frac{(\eta_{ijk} + \mu_{ij}\mu_{ik})\left((1-\mu_{ij})(1-\mu_{ik}) + \eta_{ijk}\right)}{(\mu_{ij}(1-\mu_{ik}) - \eta_{ijk})\left(\mu_{ik}(1-\mu_{ij}) - \eta_{ijk}\right)}.$$

We define the α estimating equations by

$$\sum_{i=1}^{N} C_i^T U_i^{-1}(S_i - \eta_i) = 0$$

where

$$\begin{aligned} C_i &= \partial\eta_i/\partial\alpha, \\ S_i^T &= \left(S_{i12}, \ldots, S_{in_i(n_i-1)}\right), \\ \eta_i^T &= \left(\eta_{i12}, \ldots, \eta_{in_i(n_i-1)}\right). \end{aligned}$$

Again, U_i is often taken to be I. Notice that we could also use an appropriate link function, i.e., we might assume $\ln\Omega_{ijk} = Z_{ijk}^T\alpha$ for some covariates Z_{ijk}. Putting these two sets of equations together we get

$$\sum_{i=1}^{N} \begin{pmatrix} \left(\frac{\partial\mu_i}{\partial\beta}\right)^T & 0 \\ 0 & \left(\frac{\partial\eta_i}{\partial\alpha}\right)^T \end{pmatrix} \begin{pmatrix} V_i & 0 \\ 0 & U_i \end{pmatrix} \begin{pmatrix} Y_i - \mu_i \\ S_i - \eta_i \end{pmatrix}$$

These estimating equations are called GEE-1 by Liang *et al.* (1992).

COMMENT. If our primary interest is in estimating β, then asymptotic theory tells us that it matters little how we estimate α, since the same asymptotic distribution for $\widehat{\beta}$ applies for any consistent estimate of α. With finite samples, less is known about the impact of the estimate for α.

Chapter 7

Likelihood Models for Repeated Binary Data

In Section 6.1 we showed the close correspondence between exponential family likelihood theory and quasi-likelihood for generalized linear models in the univariate setting. In the multivariate case, it is not possible to completely generalize the GLM theory to likelihoods in a way which is entirely satisfactory for non-normal data. Here we will deal with some approaches to constructing likelihood models for repeated binary data. To fix ideas, we will first consider a single sample, no covariates, with $n_i = n$. We also drop the subscript i for much of the discussion for simplicity.

With measured multivariate responses, likelihood based analyses are invariably based on the Multivariate Normal (MVN) distribution, although how the mean and variance are parameterized may differ for different models. The MVN has many attractive features:

1. In the general case, the distribution is indexed by n parameters for the mean μ and $n(n+1)/2$ parameters for the variance-covariance matrix Σ.

2. Any subset of the n-vector Y, say Y_s, also has a multivariate normal with μ_s and Σ_s being the corresponding subsets of (μ, Σ). This property is known as reproducibility.

3. The MVN ensures consistency of ML estimates of μ and Σ, *even if* MVN does not hold. In particular, if $E(Y) = X\beta$ and $V(Y) = \Sigma(\alpha)$, then the ML estimates of (β, α) will be consistent even if Y is *not* MVN.

4. The parameters μ and Σ are distinct.

In contrast, many of these features do not hold when $Y^T = (Y_1, \ldots, Y_n)$ and each Y_j is binary. One attractive feature of binary data is that it is always true that the joint pdf of Y is multinomial with 2^n points in the sample space; that is, it is not an assumption like the MVN, but can be derived from first principles.

1. In the most general case, the joint pdf has $2^n - 1$ parameters, which grows much more rapidly than $n + n(n+1)/2$ as n gets large. This implies we need models which permit parsimonious parameter specification. The parsimonious parameter models are derived by making assumptions on the multivariate distribution.

2. Given a subset Y_s, the joint distribution of the Y_s is also multinomial, but the parameters of the pdf of Y_s are now sums of the parameters of the pdf of Y. For example, if $n = 3$ and

$$\pi_{i_1 i_2 i_3} = P(Y_1 = i_1 \text{ and } Y_2 = i_2 \text{ and } Y_3 = i_3), \qquad i_1, i_2, i_3 = 0, 1,$$

denote the parameters of the cell probabilities for (Y_1, Y_2, Y_3), then the marginal distribution of (Y_1, Y_2) is again multinomial, but with parameters

$$\begin{aligned} \pi_{i_1 i_2} &= \sum_{i_3=0}^{1} \pi_{i_1 i_2 i_3} \\ &= P(Y_1 = i_1 \text{ and } Y_2 = i_2). \end{aligned}$$

3. The parameters for the mean and variance are functionally related:

$$\text{var}\,(Y_j) = E(Y_j)(1 - E(Y_j)), \qquad j = 1, \ldots, n,$$

and

$$\text{cov}\,(Y_j, Y_k) = P(Y_j = Y_k = 1) - E(Y_j)E(Y_k), \qquad j, k = 1, \ldots, n.$$

4. As we have discussed in Chapter 6, we also typically use a nonlinear link function to relate covariates to $E(Y_j) = \mu_j$.

All of these features mean that likelihood models for binary responses are more difficult to formulate than was the case with linear models in the MVN setting.

When dealing with the GLM model for multivariate data with covariates, we seek to specify a distribution for each $Y_{i\,(n\times 1)}$, $i = 1, \ldots, N$, while retaining the basic regression model of interest, namely that

$$\mu_i = E(Y_i) = g(X_i\beta) \tag{7.1}$$

for

$$\ell\,(\mu_i) = X_i\beta. \tag{7.2}$$

By definition, the distribution of Y_i is multinomial; we can view the mean restriction as imposing a model on the n one-way margins of an n dimensional array with cell probabilities $\pi_{i\,(2^n\times 1)}$. The cell probabilities depend upon i through the covariates specifying the margins. It may also be desirable to let the association parameters depend upon covariates. Thus the main issue for likelihood modeling is how to model higher order associations in the table in a flexible and interpretable way that is consistent with the model for the mean μ_i. Several general approaches have been suggested; we focus here on two approaches, both of which are related to the general log-linear modeling system for contingency tables. We will first briefly review this approach to modeling associations with multivariate binary responses, and then discuss two modifications for longitudinal data.

7.1 A brief overview of log-linear models

These models were introduced by Birch (1963), and they have enjoyed considerable success for 1) studying associations among a set of n binary (or categorical) variables or for 2) logistic regression when one of the n variables is an outcome, and the remainder are categorical predictors (Bishop, Fineberg and Holland, 1975).

To explain basic ideas, we first consider $n = 3$, and assume no covariates. Again, we generally drop the subscript i. The basic principle is that the multinomial cell probabilities are not especially useful for studying associations or the effects of covariates, so we make a transformation from π to some other parameter set which is useful. The transformation needs to be one-to-one and invertible, and we need the new parameters to have meaningful interpretations.

The log-linear transform can be written in the form

$$\underset{(2^n-1)\times 1}{\lambda} = \underset{(2^n-1)\times 2^n}{C_1^T} \ln \underset{2^n\times 1}{\pi}$$

for a certain C_1. The elements of λ can be partitioned as:

Label	**Effects**	**Number**
main effects	$\lambda_1, \lambda_2, \lambda_3, \ldots, \lambda_n$	n
$2-$ way effects	$\lambda_{12}, \lambda_{13}, \ldots, \lambda_{n(n-1)}$	$\binom{n}{2}$
$3-$ way effects	$\lambda_{123}, \lambda_{124}, \ldots, \lambda_{n(n-1)(n-2)}$	$\binom{n}{3}$
$\vdots$	$\vdots$	$\vdots$

where, for $n = 3$,

$$\lambda_1 = \ln \frac{\pi_{1**}}{\pi_{0**}},$$

and similarly, for λ_2, λ_3,

$$\lambda_{12} = \ln \frac{\pi_{11*}\pi_{00*}}{\pi_{10*}\pi_{01*}},$$

and similarly for λ_{13}, λ_{23}, and

$$\lambda_{123} = \ln \left\{ \frac{\pi_{111}\pi_{001}}{\pi_{101}\pi_{011}} \div \frac{\pi_{110}\pi_{000}}{\pi_{100}\pi_{010}} \right\},$$

and where $*$ indicates taking the geometric mean over the omitted subscripts. Thus

$$\lambda_1 = \ln \frac{\pi_{1**}}{\pi_{0**}} = \ln \pi_{1**} - \ln \pi_{0**}$$

where

$$\pi_{1**} = (\pi_{111}\pi_{110}\pi_{101}\pi_{100})^{1/4}$$

so that

$$\ln\ \pi_{1**} = \frac{1}{4}\ \ell_{1++} = \bar{\ell}_{1..}$$

where

$$\ell_{ijk} = \ln\ \pi_{ijk}.$$

It follows that

$$\lambda_1 = \bar{\ell}_{1..} - \bar{\ell}_{2...}$$

Similar expressions can be derived for λ_{12} and λ_{123}. From this it is apparent that $\lambda = C_1^T \ln \pi$ for an appropriate C_1 since each element λ is a linear combination of the elements of $\ln \pi$. Note that λ is $(2^n - 1)$ and π is 2^n, but $\sum \pi = 1$, hence there are only $2^n - 1$ unique cell probabilities. As we will discuss, the "higher order" λ's $(\lambda_{12}, \ldots, \lambda_{123})$ have interpretation as associations.

The log-linear transformation has many attractive features:

1. Apart from $\lambda_1, \ldots, \lambda_n$, which are often called the main effects, the higher order parameters can be thought of as log odds-ratios and differences of log odds-ratios, etc. To see this, consider the highest order term, or λ_{123} when $n = 3$. Rewriting λ_{123} as

$$\lambda_{123} = \ln \left\{ \frac{P(Y_1 = 1, Y_2 = 1 \mid Y_3 = 1)\, P(Y_1 = 0, Y_2 = 0 \mid Y_3 = 1)}{P(Y_1 = 1, Y_2 = 0 \mid Y_3 = 1)\, P(Y_1 = 0, Y_2 = 1 \mid Y_3 = 1)} \Bigg/ \right.$$

$$\left. \frac{P(Y_1 = 1, Y_2 = 1 \mid Y_3 = 0)\, P(Y_1 = 0, Y_2 = 0 \mid Y_3 = 0)}{P(Y_1 = 1, Y_2 = 0 \mid Y_3 = 0)\, P(Y_1 = 0, Y_2 = 1 \mid Y_3 = 1)} \right\},$$

where we use the fact that

$$P(Y_1 = i_1, Y_2 = i_2, \mid Y_3 = i_3) = \pi_{i_1 i_2 i_3} / \pi_{++i_3}.$$

Thus λ_{123} is the log odds-ratio measuring association between Y_1 and Y_2 given $Y_3 = 1$, minus the same log odds-ratio conditional on $Y_3 = 0$; i.e., $e^{\lambda_{123}}$ is a ratio of odds-ratios. Note that by symmetry, λ_{123} can also be thought of measuring odds-ratios for variables Y_1 and Y_3 given Y_2, or Y_2 and Y_3 given Y_1.

If $\lambda_{123} = 0$, it implies that these 2-way conditional odds-ratios are independent of the value of the variable being conditioned on, i.e.,

$$\text{OR}(Y_1, Y_2 \mid Y_3 = 1) = \text{OR}(Y_1, Y_2 \mid Y_3 = 0),$$

and similarly for $\text{OR}(Y_1, Y_3 \mid Y_2)$, etc. In this case, the two-way parameters λ_{12}, etc., are directly interpretable as log odds-ratios, and it follows from setting $\lambda_{123} = 0$, and using the definition of λ_{12}, etc., that

$$\begin{aligned} \lambda_{12} &= \text{OR}(Y_1, Y_2 \mid Y_3), \\ \lambda_{13} &= \text{OR}(Y_1, Y_3 \mid Y_2), \\ \lambda_{23} &= \text{OR}(Y_2, Y_3 \mid Y_1). \end{aligned}$$

2. To specify parsimonious models, we can take the higher associations (λ's) to be zero and get meaningful reduced models. The pairwise model is a popular choice:

$$\begin{aligned} \lambda &= (\lambda_1, \lambda_2, \lambda_3, \lambda_{12}, \lambda_{13}, \lambda_{23}), \\ \lambda_{123} &= 0, \end{aligned}$$

partly because this reduces us to n parameters for the mean $(\lambda_1, \lambda_2, \lambda_3)$ and $n(n-1)/2$ for the associations. With longitudinal data, we might also consider first order Markov models:

$$\lambda = (\lambda_1, \lambda_2, \ldots, \lambda_n, \lambda_{12}, \lambda_{23}, \ldots, \lambda_{n(n-1)}),$$

where

$$\lambda_{13} = \lambda_{14} = \cdots \lambda_{123} = \lambda_{234} = 0.$$

3. It is easy to characterize and compute the MLE's of λ by using Iterative Proportional Fitting (IPF).

4. The λ's are variation independent, i.e., $\lambda \in \Re^k$, where k = dim λ, so there are no restrictions on λ's, i.e., log-odds ratios do not depend on the margins $(\lambda_1, \ldots, \lambda_n)$.

5. Properties 4 and 5 are related to the fact that λ is the vector of canonical parameters in the exponential family representation of the multinomial. As shown by Cox (1972), for $n = 3$; we may write

$$\pi_{i_1 i_2 i_3} \propto \exp\left\{ \sum_j \lambda_j i_j + \sum_{j<k} \lambda_{ij} i_j i_k + \lambda_{123} i_1 i_2 i_3 \right\}.$$

The log-linear representation of π forms the basis for approaches (discussed in Section 7.3) suggested by Zhao and Prentice (1990) and Fitzmaurice and Laird (1993). The major difficulty with the log-linear model is that the "main effects," $\lambda_1, \ldots, \lambda_n$, are not very interesting or meaningful; this might be expected from a model designed to study only associations. If we are interested in $\mu_{ij} = E(Y_{ij}|X_{ij})$ as a function of covariates, this transformation is not attractive because the μ_{ij}'s are not a simple function of the λ's. We now discuss the Multivariate Logistic Transform which can be viewed as very similar to the log-linear, but the model uses marginal rather than conditional odds-ratios.

7.2 The Multivariate Logistic Transform (MLT)

The MLT is quite similar in nature to the log-linear transform. It is defined for the general case by

$$\underset{2^n \times 1}{\delta} = \underset{2^n \times 2^n}{C_2^T} \ln \underset{2^n \times 2^n}{L} \underset{2^n \times 1}{\pi}$$

for appropriate matrices C_2 and L. For $n = 3$, define

$$\delta_0 = \ln\left(\sum \pi\right) = 0,$$

$$\delta_1 = \text{logit } \mu_1 = \ln(\pi_{1++}/\pi_{0++}),$$

and similarly for δ_2 and δ_3;

$$\delta_{12} = \ln \frac{\pi_{11+}\pi_{00+}}{\pi_{10+}\pi_{01+}},$$

and similarly for δ_{13} and δ_{23}, and

$$\delta_{123} = \lambda_{123}.$$

Here δ_0 is a normalizing constant, ensuring $\sum \pi = 1$. So the MLT is exactly like the log-linear transform, except a "+" replaces a "*" in the omitted subscript. Thus we can partition δ as

Label	Parameter	Number
main effects	$\delta_1, \delta_2, \delta_3$	3
two-way associations	$\delta_{12}, \delta_{13}, \delta_{23}$	3
three-way association	δ_{123}	1.

Hence the main effects are precisely the parameters of interest, and the associations are marginal, rather than the conditional odds ratios.

The MLT was originally introduced by Grizzle, Starmer and Koch (1969), who proposed a weighted least squares analysis. McCullagh and Nelder (1989) coined the MLT phrase for $n = 3$, and proposed a maximum likelihood analysis. The general case is considered in Glonek and McCullagh (1995).

Remarks.

1. Multivariate logistic models are attractive because the main effects are those of most interest in many longitudinal data settings. As with log-linear models, we can set higher order terms to zero and get meaningful models.

2. Unlike log-linear models, δ is not variation independent, and there may be margin incompatibility.

3. As with log-linear models, there is no closed form solution for estimating δ under ML, nor, except in special cases, can one invert the transform to express π in terms of δ. This must be done iteratively, and makes computations complex.

4. The parameter sets are not orthogonal. To be explicit, suppose now that each individual has covariate vectors X_i for Y_i and F_i for the association parameters, and we write

$$\delta_i = \begin{pmatrix} \delta_{i1} \\ \vdots \\ \delta_{in} \\ \delta_i^{HO} \end{pmatrix} = \begin{pmatrix} X_i\beta \\ F_i\alpha \end{pmatrix} \begin{matrix} p \times 1 \\ m \times 1 \end{matrix}$$

   where δ_i^{HO} denotes the higher order effects (i.e., δ's with at least two subscripts) and α is a parameter indexing the associations. The information matrix, $\mathcal{I}\begin{pmatrix}\beta \\ \alpha\end{pmatrix}$, is *not* block diagonal, as it is in the MVN setting.

5. Because δ involves marginal moments of all order, it is relatively easy to handle missing responses (assuming MAR).

6. The MLT is reproducible.

7.3 A Mixed Parameter Transform

One way to retain the attractive features of both the log-linear and the MLT models is to use a mixed parameterization: Make a transformation from π to (δ^L, λ^{HO}) where $\delta^L = (\delta_1, \ldots, \delta_n)$ and λ^{HO} is λ without the main effects. Thus

$$\pi \longrightarrow \begin{pmatrix} \delta^L \\ \lambda^{HO} \end{pmatrix}$$

and given covariate matrices X_i and F_i for the ith subject we can write

$$\pi_i \rightarrow \begin{pmatrix} \delta_i^L \\ \lambda_i^{HO} \end{pmatrix} \rightarrow \begin{pmatrix} X_i\beta \\ F_i\alpha \end{pmatrix}.$$

Since the λ^{HO} are the canonical parameters, several of the attractive features of log-linear models are retained.

A general expression for the likelihood equations under the mixed model for (β, α) are given by:

$$\sum_{i=1}^{N} \begin{pmatrix} \frac{\partial\mu_i}{\partial\beta} & 0 \\ \frac{\partial\tau_i}{\partial\beta} & \frac{\partial\tau_i}{\partial\alpha} \end{pmatrix}^{-1} \begin{pmatrix} \text{var}(Y_i) & \text{cov}(Y_i, T_i) \\ & \text{var}(T_i) \end{pmatrix}^{T} \begin{pmatrix} Y_i - \mu_i \\ T_i - \tau_i \end{pmatrix} = 0,$$

where T_i is the vector of sufficient statistics for λ^{HO}: $T_i = (Y_{i1}Y_{i2}, Y_{i1}Y_{i3}, \ldots, Y_{i1}Y_{i2}\ldots Y_{in})$ and $\tau_i = E(T_i)$. If we fit a parsimonious model, some of the higher order λ's will be set to zero, and the dimension will be reduced. For the mixed model parameterization, Fitzmaurice and Laird (1993) have shown that these likelihood equations can be rewritten as

$$\sum_{i=1}^{N} \begin{pmatrix} \frac{\partial\mu_i}{\partial\beta} & 0 \\ 0 & F_i \end{pmatrix}^{T} \begin{pmatrix} \Sigma_i^{-1} & 0 \\ -\text{cov}\,(Y_i, T_i)\,\Sigma_i^{-1} & I \end{pmatrix} \begin{pmatrix} Y_i - \mu_i \\ T_i - \tau_i \end{pmatrix} = 0,$$

where $\Sigma_i^{-1} = \text{var}(Y_i)$. Thus the likelihood equations for β are identical to GEE, where $W_i = \Sigma_i^{-1}$:

$$\sum_{i=1}^{N} \left(\frac{\partial\mu_i}{\partial\beta}\right)^{T} W_i\,(Y_i - \mu_i) = 0.$$

The difference is that var (Y_i) is derived from the model for the distribution of Y_i.

In addition, Fitzmaurice and Laird (1993) show that the Fisher information matrix for (β, α) is block diagonal, thus the β and α parameters are orthogonal:

$$\begin{aligned} \text{Avar}\,(\widehat{\beta}, \widehat{\alpha}) &= \mathcal{I}^{-1}(\beta, \alpha) \\ &= \begin{bmatrix} \sum \left(\frac{\partial\mu_i}{\partial\beta}\right)^{T} \Sigma_i^{-1} \left(\frac{\partial\mu_i}{\partial\beta}\right) & 0 \\ 0 & \sum F_i^T(\Gamma_i - \xi_i\Sigma_i^{-1}\xi_i^T)F_i \end{bmatrix}^{-1} \end{aligned}$$

where $\Gamma_i = \text{var}\,(T_i)$ and $\xi_i = \text{cov}\,(Y_i, T_i)$. Thus avar $(\widehat{\beta})$ is the same as the GEE variance when $W_i = \Sigma_i^{-1}$.

By analogy with GEE, we will obtain a consistent estimate of β even if the model

$$\delta_i^{HO} = F_i\,\alpha$$

is wrong so that var (Y_i) is misspecified. In this case, however the Fisher information will not give the correct variance.

REMARKS.

- Because odds ratios are invariant to changes in the margins, the k-vector (δ^L, λ^{HO}) is variation independent in $\Re^k$.
- A computational algorithm for obtaining MLE's is given in Fitzmaurice and Laird (1993).
- A drawback to this parameterization is that because the association parameters are *conditional* rather than marginal odds ratios, this approach cannot be generalized to the arbitrary unbalanced case, and the interpretation of the association parameters is not always meaningful in the longitudinal setting, i.e., the log odds ratio expressing association between Y_1 and Y_2 conditional on the future value Y_3. A related feature is that the distribution is not reproducible, i.e., any subset of Y, say Y_s, does not have a mixed parameter model representation.

Fitzmaurice, Laird and Lipsitz (1995) discuss estimation when imbalance arises due to missing responses. Although the computations are generally straightforward, many of the nice features of the complete data case are lost. In particular, the likelihood equations become

$$\sum_{i=1}^{N} \begin{pmatrix} \frac{\partial \mu_i}{\partial \beta} & 0 \\ 0 & F_i \end{pmatrix}^T \begin{pmatrix} \Sigma_i^{-1} & 0 \\ -\xi_i \Sigma_i^{-1} & I \end{pmatrix} \begin{pmatrix} E(Y_i|Y_i^{OBS}) - \mu_i \\ E(T_i|Y_i^{OBS}) - \tau_i \end{pmatrix} = 0,$$

For β this yields

$$\sum_{i=1}^{N} \left(\frac{\partial \mu_i}{\partial \beta}\right)^T \Sigma_i^{-1} \left(E(Y_i|Y_i^{OBS}) - \mu_i\right) = 0$$

so that these are no longer the same as the GEE. In particular, $\widehat{\beta}$ will generally not be consistent for β with either MCAR or MAR data, unless the distribution of Y_i is correctly specified. This differs from ML with multivariate normal responses where estimates for β are always consistent under MCAR even when the multivariate normal is misspecified. In addition, the asymptotic covariance of (α, β) is no longer zero, and the parameters are not orthogonal. Some simulations in simple cases suggest that the bias due to model misspecification in MCAR settings is small.

Chapter 8

Random Effects Models for Repeated Binary Data

The models and methods for repeated binary data which were considered in Chapter 7 are most appropriate when the data are balanced, that is, there are n common occasions of measurement and imbalance ($n_i \neq n$ for some i) arises because of missing observations. When the unequal n_i arise because of inherently unbalanced data or because of clustered designs, the most natural approach is to consider extending the LMM using random effects to the GLM setting.

By analogy to the linear case, we assume each subject has a vector of subject-specific effects, b_i, and we add $Z_i b_i$ to the linear predictor $X_i \beta$. Letting Y_i denote the $n_i \times 1$ vector of binary outcomes, we have

$$E(Y_i \mid b_i, X_i) = \mu_i^* = g(X_i \beta^* + Z_i b_i) \tag{8.1}$$

where

$$\ell\,(\mu_i^*) = X_i \beta^* + Z_i b_i, \tag{8.2}$$

ℓ is the link function, and g the inverse link function. As before, we assume that $E(b_i) = 0$ and $\text{var}(b_i) = D$. Generally, we also assume that given b_i, the Y_{ij}'s are independent.

We use the μ_i^*, β^* notation to emphasize that μ_i^* and β^* are conditional and not marginal parameters. Recall that for μ_i, β defined in Chapters 6 and 7, we assume that

$$E(Y_i \mid X_i) = \mu_i = g(X_i \beta).$$

But here we have

$$E(Y_i \mid X_i, b_i) = \mu_i^* = g(X_i \beta^* + Z_i b_i)$$

so that

$$E(Y_i \mid X_i) = E\left(\mu_i^*\right) \neq g(X_i\beta^*)$$

for nonlinear link functions. Thus the fixed effects in this mixed effects model are not directly comparable to those in the GLM, the 'mixed' model parameterization or the GEE.

In some cases, these conditional parameters are comparable to marginal parameters. For example, with the log-link where

$$E\left\{E(Y_{ij} \mid X_{ij}, b_i)\right\} = E\left(e^{X_{ij}^T\beta^* + b_i}\right)$$

and b_i is scalar, we have

$$\begin{aligned} E(Y_{ij} \mid X_{ij}) &= e^{X_{ij}^T\beta^*}\ E(e^{b_i}) \\ &= e^{X_{ij}^T\beta^* + \ln E(e^{b_i})} \end{aligned}$$

so that $\ln E(e^{b_i})$ merely acts like a constant offset for each observation. Thus apart from the intercept, $\beta^* = \beta$.

For the logistic link function

$$E(Y_{ij} \mid X_{ij}) = E(\mu_i^*) = E\left\{\exp(X_{ij}\beta + Z_{ij}b_i)/\left(1 + \exp(X_{ij}\beta + Z_{ij}b_i)\right)\right\},$$

where expection is over the distribution of b_i. Various authors (e.g., Newhaus *et al.*, 1991, Diggle *et al.*, 1994) have shown that for the logistic link function the components of β^* are typically attenuated relative to the components of β. Because the expression for μ_i^* is an integral, in the binary setting, μ_i^* usually cannot be computed except by approximation. This makes estimation complex. We will consider two general approaches to estimating the parameters in random effects models: moment estimation and likelihood approaches. We now drop the superscript * on β and μ_i for notational simplicity, it being understood that they refer to the parameters in (8.1)–(8.2). We shall also use the following notation:

$$\begin{aligned} g'\left(X_{ij}^T\beta + Z_{ij}^T b_i\right) &= g'\left(\ell_{ij}\right) = \partial\, g/\,\partial\ell_{ij}, \\ \partial g\left(X_{ij}^T\beta + Z_{ij}^T b_i\right)\Big/\,\partial b_i &= g'\left(X_{ij}^T\beta + Z_{ij}^T b_i\right) Z_{ij}^T, \end{aligned}$$

where X_{ij}^T and Z_{ij}^T are column vectors denoting the jth rows of X_i and Z_i, respectively. Further,

$$\partial g\left(X_i\beta + Z_i b_i\right)/\,\partial\, b_i = \ \mathrm{Diag}\left\{g'\left(X_{ij}\beta + Z_{ij}b_i\right)\right\} Z_i.$$

Note that for the logistic transform, $g'(\ell_{ij}) = p_{ij}(1 - p_{ij})$ where $p_{ij} = g(\ell_{ij})$. Hence

$$\partial g\left(X_i\beta + Z_i b_i\right)/\partial b_i = R_i Z_i$$

where

$$R_i = \text{ diag } \{p_{ij}(1-p_{ij})\} = \text{ diag } \left\{g'(X_{ij}\beta + Z_{ij}b_i)\right\}.$$

8.1 GEE approach to estimating β

The basic idea here is to find the marginal means and variances of Y_i in terms of the conditional β vector and D, and use a GEE-type approach to estimation. This is attractive, because as we show in Section 8.2, full likelihood approaches are numerically difficult. We here further assume that the Y_{ij}'s given b_i are independent, so that

$$\text{var } (Y_i \mid X_{ij}, b_i) = \text{diag } \{p_{ij}(1-p_{ij})\}$$

where

$$p_{ij} = P(Y_{ij} = 1 \mid X_{ij}, b_i) = g(X_{ij}\beta + Z_{ij}b_i).$$

Provided we specify the distribution of b_i, we may in principle compute

$$E(Y_i \mid X_i) = E[g(X_i\beta + Z_i b_i)]$$

and

$$\text{var } (Y_i \mid X_i) = E[\text{diag } \{p_{ij}(1-p_{ij})\}] + \text{var } p_i.$$

Notice that because of the nonlinear link function, specifying simply the moments of b_i is not generally sufficient.

For certain link functions, and distributional assumptions on b_i, e.g., probit or log and $b_i \sim N(0, D)$, it is possible to find closed form expressions for the marginal means and variances. In other cases various approximations have been used. For a probit link and $b_i \sim N(0, D)$, Zeger *et al.* (1988) show that

$$E(Y_{ij} \mid X_i) = \Phi\left(a_p(D) X_{ij}^T \beta\right)$$

where $a_p(D) = \mid DZ_{ij}Z_{ij}^T + I \mid^{-q/2}$. For the logit link there is no corresponding closed form solution, but Johnson and Kotz (1970) show that

$$E(Y_{ij} \mid X_i) \approx \text{anti logit } \{a_\ell(D) X_{ij}^T \beta\}$$

where

$$a_\ell(D) = \mid c^2 \; DZ_{ij}Z_{ij}^T + I \mid$$

and

$$c^2 = 16\sqrt{3}/(15\;\pi).$$

Zeger *et al.* (1988) use this approximation plus Taylor series approximations for var $(Y_{ij} \mid X_i)$ to get GEE type estimating equations for β and D.

This same general approach has been used by Gilmour *et al.* (1985), who used the probit link and a normal assumption for b_i to get closed form solutions for the binomial case. They develop estimating equations for β and D similar to those given by Zeger *et al.* (1988).

A related approach (Goldstein, 1991) is to write a linear model as

$$Y_{ij} \doteq p_{ij} + e_{ij}$$

where for the binary case, the e_{ij}'s are independent with

$$\text{var}\,(e_{ij} \mid b_i) = p_{ij}(1 - p_{ij}).$$

Using a first order approximation for $g\,(X_i\beta + Z_i b_i)$ around $b_i = 0$, we can write

$$Y_i \doteq g\,(X_i\beta) + (\partial g/\partial b_i)\Big|_{b_i=0} b_i + e_i.$$

Letting R_{i0} denote var$(Y_i|b_i)$ with b_i evaluated at zero, when the logit link function is used we can approximate the variance of Y_i with

$$\text{Var}\,(Y_i|X_i) = R_{i0} + R_{i0}Z_iD\;Z_i^T R_{i0}.$$

A more general expression for the case where $g'(\ell_i) \neq g(\ell_i)\,(1 - g(\ell_i))$ is given by

$$\text{var}\,(Y_i|X_i) = R_{i0} + \Delta_{i0}Z_iDZ_i^T\Delta_{i0}$$

where Δ_{i0} is diagonal with $(g'(\ell_i))$ on the diagonal, and b_i is evaluated at zero.

Since now it is assumed that

$$E(Y_i) \doteq g(X_{ij}^T\beta),$$

it is straightforward to implement GEE to estimate β given D. Breslow and Clayton (1993) show that using Fisher scoring to solve GEE can be expressed, as in the LMCD case, as iteratively weighted least squares regression of $\tilde{Y}_i$ on X_i with weight matrix W_i being proportional to the inverse of var$(Y_i|X_i)$:

$$W_i = (\Delta_{i0}^{-1}R_{i0}\Delta_{i0}^{-1} + Z_iDZ_i)^{-1},$$

and $\tilde{Y}_i$ is the "working" variable $\tilde{Y}_i = X_i^T\beta + \Delta_{i0}^{-1}(Y_i - g(X_i^T\beta))$. Since the model for Y_i has been linearized, estimates for b_i can be taken as

$$\widehat{b}_i = DZ_i^TW_i\;(\tilde{Y}_i - X_i^T\widehat{\beta}).$$

Estimates of D may be obtained using methods discussed in Section 8.4.

8.2 Likelihood Approaches

In principle, one can generalize the ML normal theory approach, estimating β and D by marginal ML, that is, integrating out the b_i and maximizing the resulting likelihood:

$$L(\beta, D) = \Pi_{i=1}^{N} \int_{R^q} f(Y_i \mid \beta, b_i) f(b_i \mid D) db_i, \tag{8.3}$$

and using empirical Bayes for b_i:

$$\widehat{b_i} = E(b_i \mid Y_i, \beta, D) \mid_{\widehat{\beta}, \widehat{D}},$$

where β, D are evaluated at the MLE's. For the binary case, assuming independence given b_i, we have

$$f(Y_i \mid \beta, b_i) = \Pi_{j=1}^{n_i} p_{ij}^{Y_{ij}} (1 - p_{ij})^{1-Y_{ij}}$$

for

$$p_{ij} = E(Y_{ij} \mid X_{ij}, b_i) = g\left(X_{ij}^T \beta + Z_{ij}^T b_i\right).$$

What makes this approach difficult is that the integral in (8.3) does not have a closed form solution in the general case, nor do the derivatives of $L(\beta, D)$. If b_i is a scalar normal random variable, so that $Z_i^T = (1, \ldots, 1)$, then Gaussian quadrature can be used to give a very good approximation to $L(\beta, D)$ using

$$L(\beta, D) = \Pi_{i=1}^{N} \sum_{l=1}^{K} h_l(Y_i, X_i\beta, D, S_l) W_l$$

where S_l and W_l are the known mass points and weights for the $N(0,1)$ integral and depend only on the number of grid points. Here

$$h_l(Y_i, X_i\beta, D, S_l) = \Pi_{j=1}^{n_i} p_{ijl}^{Y_{ij}} (1 - p_{ijl})^{1-Y_{ij}}$$

and

$$p_{ijl} = g\left(X_{ij}^T \beta + \sqrt{D}\, S_l\right).$$

Anderson and Aitken (1985) point out that using this approach, $L(\beta, D)$ can be maximized using ordinary logistic regression, with Kn_+ responses and linear predictors $X_{ij}^T \beta + \sqrt{D} S_l, \ell = 1, \ldots, K$. This approach is easily implemented, but can be unmanageable if n_+ is large. A similar approach was used for the compound Poisson by Hinde (1982). The approach taken by most authors in this setting is to seek other approximations to $L(\beta, D)$ and its derivatives.

8.3 An Approximate Likelihood Approach: *PQL*

Penalized Quasi-Likelihood (PQL) (Green, 1987) is a method for approximate quasi-likelihood estimation with random effects. Similar approaches were proposed by Laird (1978), Stiratelli *et al.* (1984), Schall (1991) and McGilchrist and Aishett (1991). These approaches are reviewed in the paper by Breslow and Clayton (1993).

The PQL approach is more general than marginal ML, since $f(Y_i \mid \beta, b_i)$ in (8.3) is replaced by its quasi-likelihood equivalent

$$ql_i = \exp\left\{ -\sum_{j=1}^{n_i} d_i(Y_{ij}, g(X_{ij}^T\beta + Z_{ij}b_i))/2\phi \right\}$$

where $d_i(\cdot)$ is an appropriate deviance function. We restrict attention to the binary case where $\phi = 1$ and the deviance is $\ln\left(p_{ij}^{Y_{ij}}(1-p_{ij})^{1-Y_{ij}}\right)$, up to an additive constant. The PQL approach is to use Laplace's method for integral approximations. After various approximations and much simplification, the log penalized quasi-likelihood can effectively be written as

$$\begin{aligned} ql(\beta, D) &= \frac{1}{2}\sum_{i=1}^{N} \ln \mid I + Z_i^T(\Delta_i^{-1}V_i\Delta_i^{-1})^{-1}Z_iD \mid \\ &\quad -\sum_{i=1}^{N}\sum_{j=1}^{n_i} \ln\left(p_{ij}^{Y_{ij}}(1-p_{ij})^{1-Y_{ij}}\right) - \frac{1}{2}\sum_{i=1}^{N} b_i^T\ D^{-1}b_i, \end{aligned}$$

where

$$V_i = (\Delta_i^{-1}R_i\Delta_i^{-1} + Z_iDZ_i^T).$$

Assuming that the dependence of V_i on β can be ignored and D is known, approximate likelihood equations for β and estimating equations for the b_i can be obtained by jointly maximizing the last two terms, i.e., maximizing

$$\sum_{i=1}^{N}\left\{\sum_{j=1}^{n_i} \ln\left(p_{ij}^{Y_{ij}}(1-p_{ij})^{1-Y_{ij}}\right) - \frac{1}{2}\Sigma b_i^T D^{-1} b_i\right\}, \qquad (8.4)$$

as a function of β and $B^T = (b_1^T, \ldots, b_N^T)$, while D is held constant. For the binary case with logit link function, this leads to

$$\sum_{i=1}^{N} X_i^T\ (Y_i - \widehat{p}_i) = 0$$

and

$$\widehat{b}_i = DZ_i^T(Y_i - \widehat{p}_i)$$

where

$$\widehat{p}_i = g(X_i^T\widehat{\beta} + Z_i\widehat{b}_i).$$

Precisely this same set of estimating equations for (β, b) was derived by Stiratelli *et al.* (1984) (see also Knuiman and Laird, 1988), using an empirical Bayes approach. Drawing on the normal analogy, the unified empirical Bayes approach to estimating β, B and D is to assign β a flat prior, estimate (β, B) by their joint posterior means, and var $(\widehat{\beta}, \widehat{B})$ by their posterior variance matrix, holding D fixed. Then D is estimated by maximizing the marginal likelihood, obtained by integrating β out of (8.3) as well. Notice that with multivariate normality, the joint posterior moments are equal to marginal posterior moments for (β, D), the posterior means equal the posterior modes and the posterior variances will be the inverse second derivative matrices.

Because of the intractability of the posteriors, posterior modes rather than means are used. But, with a flat prior for β, the posterior mode for (β, B) given D is obtained by maximizing (8.4), i.e. the empirical Bayes and PQL estimates for β, B coincide. Notice that the likelihood equations for β look exactly like ordinary logistic regression, except that the linear predictor is $X_i^T\beta + Z_i\widehat{b}_i$, not $X_i^T\beta$. When $D \doteq 0$, $\widehat{b}_i \doteq 0$ and $\widehat{\beta}$ is approximately the ordinary logistic regression estimator, since $D \doteq 0 \Rightarrow b_i \doteq 0$. Alternately, when D is very large so that $D^{-1}\widehat{b}_i \doteq 0$, the estimate for $\widehat{b}_i$ requires that

$$Z_i^T(Y_i - \widehat{p}_i) \doteq 0$$

which would be the same as treating the b_i's as fixed constants and maximizing

$$\sum_{i=1}^{N}\sum_{j=1}^{n_i} \ln\left\{p_{ij}^{Y_{ij}}(1 - p_{ij})^{1-Y_{ij})}\right\}$$

as a function of (β, B). Breslow and Clayton (1993) suggest estimating var$(\widehat{\beta})$ by $\widehat{V}$, where $\widehat{V} = (\Sigma X_i^T W_i X_i)^{-1}$.

Now consider estimation of D. In the normal case, the variance components may be found using the profile likelihood approach. Here estimating D is more complicated because of the dependence of W_i on $(\widehat{\beta}(\theta), \widehat{B}(\theta))$ where θ denotes the parameters in D. If we ignore that dependence, and replace $f(Y_{ij} \mid p_{ij})$ by a standard kernel for $(\tilde{Y}_{ij} -$

$X_{ij}^T\widehat{\beta})$, one can again obtain a similar set of scoring equations for D by maximizing

$$\frac{1}{2}\sum_{i=1}^{N} \ln \mid W_i \mid -\frac{1}{2}\sum_{i=1}^{N}(\tilde{Y}_i - X_i\widehat{\beta})^T W_i(\tilde{Y}_i - X_i\widehat{\beta})$$

for maximum likelihood, or

$$\frac{1}{2}\sum_{i=1}^{N}\ln \mid W_i \mid -\frac{1}{2}\ln \mid \sum_{i=1}^{N} X_i^T \widehat{V}^{-1} X_i \mid -\frac{1}{2}\sum_{i=1}^{N}(\tilde{Y}_i - X_i\widehat{\beta})^T W_i(\tilde{Y}_i - X_i\widehat{\beta})$$

for REML.

The alternative approximation of Stiratelli *et al.* (1984) is to maximize the marginal likelihood to estimate D, integrating out both β and B. They avoid direct approximation of this integrated likelihood by approximating the derivatives instead. If the b_i's were observed, $\Sigma b_i b_i^T$ would be the sufficient statistic for D. Hence, as in the normal theory case, EM estimating equations yield

$$\begin{aligned}\widehat{D} = \Bigg[&\sum_{i=1}^{N} E\left((b_i \mid Y_i, \widehat{D})(E(b_i^T \mid Y_i, \widehat{D})\right) \\ &+E\left(\text{var}(b_i \mid Y_i, \beta, \widehat{D})\right) + \text{var }E(b_i \mid Y_i, \beta, \widehat{D})\Bigg] \Big/ N.\end{aligned}$$

These expectations can be readily evaluated by assuming that the joint posterior of (β, B) is approximately normal, with mean given by $(\widehat{\beta}(D)$, $\widehat{B}(D))$ and variance given by the inverse second derivative matrix. These approximations should work well provided q is small relative to each n_i and N is large, but may be poor otherwise.

The PQL approach for estimating θ and β gives biased results, and in some cases the bias can be substantial. Breslow and Lin (1995) and Lin and Breslow (1996) study the bias and give simple correction formulas for using the PQL approach when $q = 1$.

Chapter 9

Nonignorable Nonresponse

The techniques discussed in the preceding sections all can accommodate missing data and/or unbalanced designs, but valid inferences require assumptions on the missing data mechanism (MDM). The GEE methodology requires missing completely at random (MCAR); the likelihood approaches are valid with missing at random (MAR) nonresponse mechanisms, but then further require that the likelihood be correctly specified. We note that the GEE can be adapted to handle MAR mechanisms as well by using nonresponse weights, similar to nonresponse adjustments used in sample surveys (Robins, Rotnitzky and Zhao, 1995).

If the MDM depends upon the unobserved responses, given the observed responses, it is said to be nonignorable. Then the inferential issues are more complex, because here we need to make unverifiable assumptions on the MDM in order to have valid inferences. We consider some examples, many of which are univariate since the issues are not fundamentally different for the multivariate case.

Example 1. In estimating mean income (or a regression of income on covariates), nonresponse is often assumed to be more prevalent among those with very high or very low incomes (Greenlees, *et al.*, 1982).

Example 2. Studies of factors which influence wages in workers may be biased by the omission of workers who are not currently in the workforce due to unemployment (Heckman, 1976). Heckman used the term selection bias to describe such self-selected samples and noted the connection to nonignorable nonresponse. He proposed methods for dealing with nonresponse in the univariate setting based on estimating the parameters of a model for $P(\text{missing})$. Such approaches are generally referred to as selection model approaches.

Example 3. Many voters respond "don't know," "undecided," or "will not vote" in pre-election surveys. Such respondents are usually ignored and predictions are based on respondents who give a preference. However, voter unwillingness to commit to an answer may be nondifferential depending upon who they ultimately vote for. Baker and Laird (1988) reanalyze the 1948 Truman–Dewey preelection polls which were notorious for failing to predict a win for Truman. Their analysis suggests that voters in the "don't know" or "undecided" categories were far more likely to favor Truman; reanalysis under this scenario predicts an outcome very close to the election and to the voter exit polls.

We now turn to some longitudinal data examples.

Example 4. In cancer chemotherapy trials, increasing emphasis is placed on evaluating quality-of-life of patients on different therapies. Quality-of-life is typically measured quarterly with a self-report questionnaire. The questionnaire is long and there is considerable missing data even among those patients who remain on the study protocol. Cancer chemotherapy patients often have periods of debilitating pain or illness, and a plausible assumption is that these patients would be less likely to make the effort to respond under these circumstances. Nonresponse at a particular occasion is less likely to be predicted by quality-of-life score at the previous occasion (which might be observed), but more likely depends on current unobserved score.

Other examples of intermittent nonignorable response in longitudinal surveys are more difficult to find. A more common problem in this setting is monotone missingness, or dropouts, when subjects are removed from study, dropout or otherwise become unavailable for study at some point. If P(dropout at t) depends only on observed covariates and past history, i.e., y_{ij}'s observed prior to dropout, then dropout is ignorable. If P(dropout at t) depends upon the unobserved response which would have been obtained (say Y_{it}) had the subject not dropped out, dropout is nonignorable. Several examples can be found in Diggle and Kenward (1994).

Example 5. The U.S. government uses an annual panel survey of doctoral level scientists and engineers to estimate current unemployment rates among scientists and engineers. A follow-up of those ceasing to participate in the survey revealed that a disproportionate number had moved abroad, relative to those remaining in the survey. Since the government is only interested in employment in the U.S., this is a clear example of nonignorable dropout.

Example 6. In many clinical trials, patients can be removed from the protocol for reasons related to the outcome of interest. For example, in a drug trial of psychosis in schizophrenics, the endpoint was reduction in psychosis after six weeks on drug (Hogan and Laird, 1997a). But many patients were removed from study drug early due to lack of effect. In cases such as this, one might argue that dropout is ignorable since it can be predicted by the observed prior outcomes, although the indications used by clinicians to remove patients from a study protocol may not be quantified in the data available for analysis.

9.1 Terminology

Let us assume that for each subject, the vector of complete responses, Y_i can be partitioned into $Y_i^{\mathrm{OBS}} = (Y_{i1}, \ldots, Y_{ik})^T$ and $Y_i^{\mathrm{MIS}} = (Y_{ik+1}, \ldots, Y_{in})^T$, and d_i denotes a subject's dropout time where $t_k < d_i \leq t_{k+1}$. This simple model implies Y_{ij} is always observed prior to d_i and never after. Of course, if $d_i > t_n$, subjects do not drop out.

Nonignorable Nonresponse:

If

$$P(t_k < d_i \leq t_{k+1} \mid Y_i, \psi) = P\left(t_t < d_i \leq t_{k+1} \mid Y_i^{\mathrm{OBS}}, Y_i^{\mathrm{MIS}}, \psi\right), \quad (9.1)$$

i.e., the probability of nonresponse due to dropout depends on future unobserved values Y_i^{MIS}, even conditional on the past, Y_i^{OBS}, then dropout is nonignorable. The term nonignorable comes from the fact that valid likelihood based inferences require specification of the nonresponse mechanism (which is also called the selection model), and maximizing the observed data likelihood of $(Y_i^{\mathrm{OBS}}, d_i)$. This will be described in Section 9.2.

The term informative dropouts is often used in this setting. We define **non**-informative dropout to mean:

Noninformative Dropout:

$$f\left(Y_i^{\mathrm{MIS}} \mid X_i, Y_i^{\mathrm{OBS}}, d_i\right) = f\left(Y_i^{\mathrm{MIS}} \mid X_i, Y_i^{\mathrm{OBS}}\right) \quad (9.2)$$

where the same partitioning of $Y_i^{\mathrm{MIS}}, Y_i^{\mathrm{OBS}}$ is used in both sides of (9.2), and determined by $t_k < d_i \leq t_{k+1}$. Noninformative thus implies that the **fact** of dropout implies no new information about future values that is not already contained in a patient's history, $(X_i, Y_i^{\mathrm{OBS}})$. Conversely, if equality does **not** hold in (9.2), then dropout is said to be informative, i.e., the fact of dropping out is informative about future (missing) values, even conditional on past history.

It is straightforward to see that ignorable dropout and noninformative are equivalent concepts, and informative and nonignorable are as well. An advantage of the informative (noninformative) representation is that it makes it clear what is inherently inestimable in this setting, and what features of the model the results will be sensitive to. The basic problem is that without additional assumptions, there are no data on those observed after d_i to estimate responses for an individual with dropout d_i.

9.2 Methodology: General Comments

There are two general model-based approaches to handling nonignorable nonresponse (Hogan and Laird, 1997b): selection-modeling and mixture-modeling. In this section we give a brief overview of each; detailed case studies are given in Section 9.3.

Selection Modeling. As noted earlier, the term selection modeling refers to approaches which specify both the distribution of Y_i and R_i (or d_i) and base inference on the joint distribution. The contribution of the ith subject to the likelihood can be expressed as

$$L_i(\beta, \theta, \psi) = \int f(Y_i \mid \beta, \theta)\ p(d_i \mid Y_i, \psi) dY_i^{\mathrm{MIS}}.$$

Since $p(d_i \mid Y_i, \psi)$ depends upon the unobserved Y_i's, estimation of the model parameters is at best difficult, and in general sensitive to model specification. To see why, consider the simple univariate case where we assume that $Y_i \sim N(\beta, \theta)$ and p (nonresponse) is monotone in Y_i. Then intuitively, if nonresponse is independent of Y_i, our sample histogram using the observed data should look approximately normal; if P(nonresponse) is monotone increasing in Y_i, it will be skewed the left, and right skewed if P(nonresponse) is decreasing in Y_i:

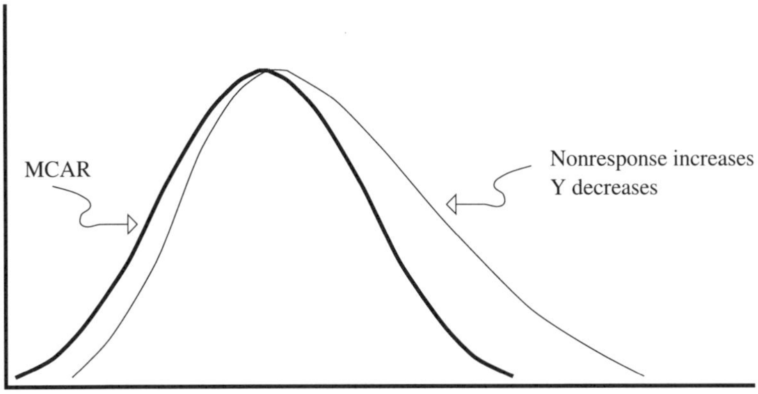

FIGURE 9.1.

Thus estimation of P(nonresponse) relies heavily on assumed normality. If we assume log-normality for Y_i, we could see quite different results.

In the case with covariates, multivariate responses and arbitrary patterns of missingness, it is generally not possible to permit all possible models because of nonidentifiability problems. In fact, nonidentifiability can arise in the simple response setting where $n = 2$, each Y_{ij} is dichotomous, and there is missingness in Y_{i2} only. In this case, one must either impose restrictions on $f(Y_i)$ or on $f(R_i \mid Y_i)$ where R_i is an indicator of missing Y_{i2}, to estimate the model parameters. Little and Rubin (1989, Chapter 11) show that the following models are **not** identifiable:

(1) $$f(Y_i) = f_1(Y_{i1})f_2(Y_{i2}) \quad \text{and} \quad f(R_i \mid Y_i) = f(R_i \mid Y_{i2}),$$

(2) $$f(Y_i, R_i) = f_1(Y_{i1} \mid R_i)f_2(Y_{i2} \mid R_i)f(R_i),$$

(3) $f(Y_i, R_i)$ completely general,

and, in addition, a model which specifies that all two-way associations are non zero: $(Y_{i1}, R_i), (Y_{i2}, R_i)$ and (Y_{i1}, Y_{i2}). Thus the only possible nonignorable model in this setting is

(4) $$f(Y_i) \text{ arbitrary} \quad \text{and} \quad f(R_i \mid Y_i) = f(R_i \mid Y_{i2}).$$

Applications of selection modeling in this setting typically presume a simplified model for dropout or nonresponse, partly in order to overcome technical limitations, partly to ensure identifiability and partly to have reasonable assumptions about the MDM. In the dropout (monotone missing) setting a typical assumption is P(dropout) depends only on current unobserved value. Diggle and Kenward (1994) and Troxel *et al.* (1998) assume missingness depends only on current unobserved value.

Random Effects Selection Models

A somewhat different approach, which is especially suitable for longitudinal data, is random effects selection modeling. Here we allow P(dropout) to be a function of unobserved random effects. Recall the random effects model

$$Y_i = X_i\beta + Z_i b_i + e_i,$$

where as usual, we take b_i and e_i to be independent $N_q(0, D)$ and $N_n(0, \sigma^2 I)$. We now further assume that d_i has a distribution which depends upon b_i. For example, suppose we are fitting linear models, so that

$$Z_i = \begin{bmatrix} 1 & t_1 \\ \vdots & \vdots \\ 1 & t_n \end{bmatrix}$$

and (b_{0i}, b_{1i}) represent the ith individual's unobserved, random, residual slope and intercept. If we assume that

$$f(d_i \mid Y_i, b_i, \psi) = f(d_i \mid b_i, \psi),$$

i.e., dropout given b_i does not depend upon Y_i, then we still have non-ignorable dropout since $f(d_i \mid Y_i^{\text{OBS}}, Y_i^{\text{MIS}}, \psi)$ will in general depend on Y_i^{MIS}.

An attractive feature of this model is that it follows immediately that

$$\underset{(n_i \times 1)}{Y_i^{\text{OBS}}} = \underset{(n_i \times p)(p \times 1)}{X_i^{\text{OBS}}\beta} + \underset{(n_i \times q)(q \times 1)}{Z_i^{\text{OBS}}b_i} + \underset{(n_i \times 1)}{e_i}$$

and the marginal distribution of (Y_i^{OBS}, d_i) is obtained as

$$f(Y_i^{\text{OBS}}, d_i) = \int_{R^q} f(Y_i^{\text{OBS}} \mid b_i, X_i, \beta, \theta)\ f(b_i \mid D)\ f(d_i \mid b_i, \psi) db_i.$$

This model was originally proposed by Wu and Carroll (1988) who assumed that the probit of the probability of dropout at each occasion, conditional on being at risk, was linear in b_{1i} and an indicator of period. The probit model allows a closed form expression for the integral, in terms of the cumulative normal, but the complex nature of the computations led them to seek alternative approximations.

Schluchter (1992) and DeGruttola and Tu (1994) independently noticed that if d_i (or some transformation of d_i) is $N_1(u_i^T\psi_0 + b_i^T\psi_1, \tau^2)$

where u_i is a vector of covariates, then the marginal distribution of (Y_i^{OBS}, d_i) is easily obtained as

$$\begin{pmatrix} Y_i^{\text{OBS}} \\ d_i \end{pmatrix} = N\left[\begin{pmatrix} X_i^{\text{OBS}}\beta \\ u_i^T\psi_0 \end{pmatrix} \begin{pmatrix} Z_i^{\text{OBS}} D Z_i^{\text{OBS}^T} + \sigma^2 I & Z_i^{\text{OBS}} D\psi_1 \\ \psi_1^T D Z_i^{\text{OBS}^T} & \psi_1^T D\psi_1 + \tau^2 \end{pmatrix}\right].$$

The marginal distribution for Y_i^{OBS} is the same as the usual random effects distribution, but d_i has a factor-analysis structure rather than a random effects structure. That is, the model for Y_i^{OBS} resembles ordinary random effects, while that for d_i resembles factor analysis. If there is no censoring of the d_i's then the analysis is relatively straightforward. In many cases, however, d_i may be censored. In the case of informative dropouts in clinical trials, d_i is time to dropout due to endpoint related reasons; hence noninformative dropouts (e.g., withdrawals due to non informative reasons) and completers will be censored.

Notice that this particular random effects selection model also implies that

$$E(Y_i^{\text{OBS}} \mid X_i, d_i) = X_i^{\text{OBS}}\beta + Z_i^{\text{OBS}} D\psi_1(d_i - u_i^T\psi_0)/(\psi_1^T D\psi_1 + \tau^2),$$

i.e., the conditional mean of Y_i^{OBS} depends linearly upon d_i. From the joint distribution of (Y_i, b_i, d_i), it is also straightforward to see that the conditional mean of b_i given d_i is likewise linear in d_i. Wu and Bailey (1988, 1989) noted that the model

$$E(b_i \mid d_i) \doteq \tilde{\psi}_0 + \tilde{\psi}_1 d_i$$

holds approximately for a broad class of random effects section models even when the normal assumption on d_i does not hold. More generally, Follman and Wu (1995) term these random effects models as shared parameters models, and have extended the methodology to the generalized linear model setting.

Mixture-Models. As we have seen, with random effects selection models the conditioning $f(d_i \mid Y_i)$ can be reversed to $f(Y_i \mid d_i)$; this concept underlies mixture models, which index the joint distribution of (Y_i, d_i) by $f(Y_i \mid d_i)\ f(d_i)$. This is in contrast to selection models which index the joint distribution by $f(Y_i)f(d_i \mid Y_i)$. Because the target of inference is ordinarily the parameters in $f(Y_i)$, the selection models seems more intuititive to many statisticians. In addition, the dependence of Y_i on d_i in $f(Y_i \mid d_i)$ is difficult to appreciate in a causal setting since it implies early values of Y_i may depend on subsequent dropout. Perhaps the best way to think about the mixture model is simply as an alternative method for factoring $f(Y_i, d_i)$ which offers some advantages over the selection model. If $f(Y_i \mid d_i) = f(Y_i^{\text{MIS}} \mid Y_i^{\text{OBS}}) f(Y_i^{\text{OBS}} \mid d_i)$ then the missingness is non-informative. In the general case where $f(Y_i^{\text{MIS}} | Y_i^{\text{OBS}})$ depends on d_i as well, missingness is informative. When inference is desired for $f(Y_i)$ we must "mix" over the distribution of d_i, giving

$$f(Y_i) = \int f(Y_i \mid d_i)\, f(d_i) dd_i,$$

hence the term mixture models.

Random effects mixture models have been considered by Wu and Bailey (1988), Hogan and Laird (1997a) and Mori *et al.* (1992). These models offer some advantages over selection models, in that simpler estimation approaches are available, even with censored observations. For example, $f(d_i)$ may be estimated using the Kaplan-Meier estimator, and $f(Y_i^{\text{OBS}} \mid d_i, X_i)$ may be estimated from those with observed failure times, provided the censoring of d_i is unrelated to Y_i^{OBS}. Since here censoring means noninformative dropout **or** completers, this condition holds. With no censoring, estimation is very straightforward and can be done using standard techniques for longitudinal data analysis once one specifies a model for $f(Y_i \mid X_i, d_i, \beta, \theta)$. See Hogan and Laird (1997b) for details.

Little (1993) gives a general method for analyzing multivariate data with missing values which is appropriate when one can classify subjects on patterns of nonresponse, R_i. Here again, models for $f(Y_i \mid R_i = r_i)$ are posited, and constraints on $P(R_i | Y_i)$ are used in order to make the models identifiable. Little (1993) calls these pattern mixture models since potentially there is a different pattern for $f(Y_i)$ given each nonresponse pattern. A general review of mixture and selection models for dropouts in given in Little (1995).

9.3 Examples

In this section we will illustrate the use of selection models and mixture models to analyze non-ignorable data. Both examples are drawn from the literature (Fitzmaurice *et al.*, 1996, and Fitzmaurice and Laird, 2000); the reader is referred to these articles for more detail.

A Selection Model. This example involves bivariate repeated measures rather than longitudinal data, but the basic methodology is the same.

The data come from a sample survey of mental health among children. Many variables were assessed, including a measure of the child's emotional or internalizing behavior. This is measured by interview from an appropriate informant using a standardized scale, and then dichotomized for analysis. It is common in mental health surveys, especially of children, to use two or more informants. In this case, the identical information was gathered from both a parent and a teacher, hence the bivariate dichotomous outcome.

Because children could not participate without parental consent, there was little missing data in the parent assessment. However, there was considerable missing data in the teacher response. Many parents refused to give permission for the teacher assessment. In addition, many teachers simply failed to respond even when permission was given. There were 2,501 parents who returned valid questionnaires, but teacher assessments were not available for 43% of these. The investigators hypothesized that non-response was non-ignorable and that the cases of externalizing behavior were under represented among the observed teacher responses. This belief stemmed in part because some parents who refused permission for the teacher interview did so because they were concerned that the teacher would give the child "bad" ratings. This is of course not clear evidence of non-ignorable nonresponse, but does provide a rationale.

We first consider the analysis if we had complete data in order to specify a data model. One objective is simply to estimate the overall prevalence based on both the teacher and parent ratings. Secondly, we want to study the relationship of covariates to the behavior ratings, and the interaction between covariates and informant on ratings, adjusting for nonresponse. Here we will consider just two covariates, sex and a dichotomized measure of dissatisfaction with family life. These questions can be addressed by using a bivariate logistic regression model, with indicators for informant, the covariates and their interactions. The data for each subject can be thought of as arising from a 2×2 cross classification of each child on the two responses, and is thus multinomial. The basic model parameterizes the three independent parameters of the multinomial into two parameters corresponding to two the marginal log-odds and one corresponding to the odds ratio.

For example, consider the model which has main effects for informants and both covariates:

$$\text{logit}P(Y_{ij} = 1) = \beta_0 + \beta_1 I_j + \beta_2 X_{1i} + \beta_3 X_{2i}$$

where Y_{ij} denotes the response of the jth informant for the ith child, $I_1 = 1$ (for parent informant), and $I_2 = 0$ (for teacher informant), and

X_{1i} and X_{2i} denote the sex, and dissatisfaction variable for the ith child.

This model implies that there are informant differences in prevalence, as well as gender and dissatisfaction effects on prevalence, but that the informant effects do not depend on covariates. For simplicity, we assume that the OR is constant, i.e., not dependent on the covariates. This completes the specification of the data model. The model so defined, as well as ones including interaction terms, is fit using maximum likelihood.

To accommodate the missing teacher responses, we also use a logistic model for the response, non-response indicator, say, $R = 1$ if observed, and zero otherwise. The model can include as predictors the parent report, the potentially unobserved parent report, the covariates given in the data model as well as possibly other covariates of nonresponse, and their interactions. If the non-response model includes the teacher observation as a predictor, then the model is nonignorable, because non-response depends upon the potentially unobserved teacher response. If it includes only the parent observation and not the teacher, the model is MAR, and if it includes neither it is MCAR. Thus an advantage of this modeling approach is that different assumptions from the Rubin (1976) missing data paradigm can easily be fit and tested.

However, it is not possible to fit completely saturated models, because then the resulting estimation problem is ill conditioned, i.e., the parameters cannot be uniquely identified from the observed data. To see this, consider the case where there are no covariates in any model. Assume further that we fit a model for nonresponse that includes both parental and teacher observations, and their interaction and a data model that has an intercept and the informant effect. In this setting, the three parameters of the data model plus the four parameters of the nonresponse model completely determine the probabilities underlying the $2 \times 2 \times 2$ cross-classification of parent observation, teacher observation, and response indicator. However, we do not directly observe this $2 \times 2 \times 2$ table; rather, we observe the 2×2 table for parent and teacher observation given $R = 1$, and we observe the parent margin of the same 2×2 table for $R = 0$. Thus the saturated model is clearly not identifiable with the observed data.

One strategy which is possible with count data is to only fit models with degrees-of-freedom equal to directly observed cells; another is to evaluate the rank of the information matrix to determine estimability. Neither of these approaches work reliably but Glonek (1999) gives a straightforward approach to determining model identifiability for the categorical data setting.

Fitzmaurice *et al.* (1996) fit by ML a single data model with four different nonresponse models to the data. The data model assumes main effects for informant, gender and dissatisfaction, and the informant by dissatisfaction interaction; the nonresponse models all include the main effects of the covariates. The MCAR model includes no effects for parent or teacher observations. The MAR model includes parent observation. The third model replaces parent observation with teacher observation and the final model includes main effect for both parent and teacher observation. The MCAR and MAR models give identical results for the data model, as expected. Comparing the log-likelihoods suggests no evidence for preferring MAR to MCAR, and the estimated prevalence of internalizing disturbance is nearly the same for teachers and parents, and for the teacher complete cases (0.189).

The results for the MCAR and the two nonignorable models are given Table 1.

There is not much variation in the estimated coefficients of the data model across any of the nonresponse models; probably because there is little evidence from any model that nonresponse depends either on parent or teacher observation. Note that the large standard errors for the coefficient of teacher response in both of the nonignorable nonresponse models indicate that there is little information in the data about this coefficient. In addition, the standard errors for the data model are also very large when both teacher and parent observations are included in the model. The Fisher Information matrix for this model, evaluated at selected parameter values about the maximum was always non-singular, however.

In this example, the overall conclusions about prevalence and the effects of covariates on prevalence vary little under different assumptions on the nonresponse mechanism. It is important to keep in mind that the class of nonignorable models considered was limited, and using other models might show a bigger impact. Despite that, the results do not give credence to the hypothesis that the missingness was nonignorable. The models fit were kept simple to avoid identifiability problems, but even so, the estimation of the non-response model and also of the data model showed large levels of uncertainty.

Table 1

Measurement Parameters			Nonresponse Parameters		
Model/Parameter	Estimate	SE	Parameter	Estimate	SE
MAR					
Intercept	-1.472	.106	Intercept	.266	.071
Informant	-.280	.125	Y_P	.105	.106
Dissat	.417	.136	Dissat	.086	.083
Gender	-.411	.086	Gender	-.075	.081
Informant × Dissat	.549	.166			
Ln(OR)	.648	.156			
-2 Log-Likelihood = 7,097.84					
GOF: $X^2 = 7.31$, 10 df, ($p \approx .70$)					
NI					
Intercept	-1.927	.515	Intercept	.132	.160
Informant	.176	.521	Y_T	1.498	2.819
Dissat	.437	.133	Dissat	.042	.101
Gender	-.414	.086	Gender	-.045	.100
Informant × Dissat	.529	.163			
Ln(OR)	.646	.156			
-2 Log-Likelihood = 7,097.58					
GOF: $X^2 = 7.04$, 10 df, ($p \approx .72$)					
NI					
Intercept	-1.966	2.223	Intercept	.124	.539
Informant	.215	2.224	Y_T	1.733	14.963
Dissat	.438	.132	Y_P	-.010	.476
Gender	-.413	.086	Dissat	.039	.200
Informant × Dissat	.528	.163	Gender	-.044	.217
Ln(OR)	.643	.228			
-2 Log-Likelihood = 7,097.57					
GOF: $X^2 = 7.04$, 9 df, ($p \approx .63$)					

A Mixture Model. Here we use data from a clinical trial on contraception to illustrate the use of mixture models to handle nonignoragle nonresponse. In the trial, two dose levels of a drug were compared. The design called for injections at baseline and every three months for one year. The primary outcome of interest was the presence or absence of a side effect, amenorrhea, during the previous three months. The vector of outcomes for a person who completes the trial is a 4×1 vector of indicators of amenorrhea (Y_i). The trial objective was to compare trends in prevalence of amenorrhea in the two groups. However, there were many subjects who failed to return for subsequent scheduled injections, and it was thought that this might be related to the presence of amenorrhea (which is not observed) during the previous period. We describe an analysis presented in Fitzmaurice and Laird (2000).

With the mixture model, we need to specify a model for how the outcomes (Y_i) depend upon treatment group, and time of dropout, and another model for time of dropout. Since there are only three dropout times (excluding completing), and the dropout times are observed for everyone, it is easy to simply estimate the dropout probabilities for each time, stratified by treatment group, using the sample proportions. Denote these probabilities by π_{ik} where i indexes treatment group and k is time of dropout, $k = 1, \ldots, 4$, the last value indicating a completer, and $\pi_{i+} = 1$, $i = 1, 2$.

For the outcome model, we could again contemplate estimating $P(Y_{ij})$ as a function of time and treatment group, stratifying on dropout status. But it is immediately obvious that there are no data to estimate $P(Y_{i4})$ if dropout occurs on the third occasion, and similarly for Y_{i3}, Y_{i4} with dropout at the second occasion, etc. Another issue, which also arises with the selection model, is that if only a few dropouts occur, then even the identifiable parameters will be poorly estimated. A solution to both of these problems is to fit simpler model which indicates how outcome depends on dropout, and condition rather than stratify on dropout.

Fitzmaurice and Laird (2000) fit several models to investigate the sensitivity of results to modeling assumption. The first model assumes that logit$(P(Y_{ij} = 1))$ follows a quadratic in time, with different time coefficients for each dose group, and with intercept depending upon time of dropout (modeled as a dummy variable). That is, the effect of dropout is to shift up or down the overall level of amenorrhea, but not to change the shape. Thus effectively, the shape of the curve for early dropouts is extrapolated from the shape which is fit to the later dropouts and the completers. A more complex model allowed the shape to depend on both dropout status and treatment group, but was assumed to be the same for all dropouts (within a treatment group) at times 1, 2 and 3. They compared the results of these models with an ignorable model where logit$P(Y_{ij} = 1)$ is just quadratic in time, with different coefficients for each dose group. In principle, a complete likelihood could be specified for the Y_{ij}'s and ML estimation used for the regression parameters and any association parameters, but they used GEE to estimate the regression models and the empirical variance for the estimated standard errors.

The final step is to obtain estimates of the marginal $P(Y_{ij} = 1) = \mu_{ij}$, conditional on dose but not dropout. This is simply a summation,

$$\widehat{\mu}_{ij} = \sum_k \widehat{P}(Y_{ij} = 1 \mid \text{time, treatment, dropout} = k)\widehat{\pi}_{ik},$$

where the estimated probabilities of $P(Y_{ij} = 1)$ are the anti-logits from the fitted regression models.

The results of model fitting are given below for the difference in mean vectors, say $\delta_j = \mu_{hj} - \mu_{lj}$ for the high and low groups at time point j. There is very little variation in the estimated δ_j, for any of the three models, except time 4 where the most complicated model estimates half the difference of the other two, but no model suggests that the treatment groups are different at the end of the trial.

Generalized Linear Mixture Models

Marginal rates of amenorrhea at the four study time pointime under three different modelling assumptions about dropout

Model	Time j	$\widehat{\delta}$ Difference	S.E.	Z	p
MCAR	1	0.017	0.023	0.73	0.4629
	2	0.089	0.025	3.54	0.0004
	3	0.109	0.030	3.65	0.0003
	4	0.052	0.036	1.46	0.1437
Dropout affects level	1	0.016	0.023	0.72	0.4734
	2	0.091	0.026	3.57	0.0004
	3	0.112	0.030	3.71	0.0002
	4	0.053	0.035	1.52	0.1275
Dropout affects slope	1	0.016	0.023	0.70	0.4864
	2	0.100	0.029	3.39	0.0007
	3	0.105	0.041	2.55	0.0108
	4	0.025	0.056	0.44	0.6614

Summary. With both the selection model and the mixture model examples, the use of several different models which reflect different assumptions about dropout are used to explore the sensitivity of results to model assumptions. We emphasize that this is perhaps the best use of the various methods available for handling nonignorable dropouts. In both of our examples, investigators felt that the nonresponse process was nonignorable, but in neither case did our models give evidence of that. Bear in mind that there are many possible models that one can fit, however, and some may well give different results. An alternative to estimating the model parameters would be to fix them at prespecified values. This has the advantage of ameliorating the problems of small number of dropouts and lack of identifiably.

Bibliography

AIDS CLINICAL TRIALS GROUP (1993). A randomized blinded trial to evaluate the safety and tolerance of high versus low dose zidovudine administered to children with human immunodeficiency virus, Protocol 128, Version 6.0 (final). National Institute of Allergy and Infectious Diseases, Washington, DC.

AMEMIYA, T. (1973). Regression analysis when the dependent variable is truncated normal. *Econometrika* **41** 997–1016.

AMEMIYA, T. (1985). *Advanced Econometrics.* Cambridge University Press.

ANDERSON, D. A. and AITKIN, M. (1985). Variance component models with binary response: interviewer variability. *J. Roy. Statist. Soc. Ser. B* **47** 203–210.

BAKER, S.G. (1995). Marginal regression models for repeated binary data with outcome subject to non-ignorable non-response. *Biometrics* **51** 1042–1052.

BAKER, S. G. and LAIRD, N. M. (1988). Regression analysis with categorical data subject to nonignorable nonresponse. *J. Amer. Statist. Assoc.* **83** 62–69.

BARNDORFF-NIELSEN, D. E. and COX, D.R. (1994) *Inference and Asymptotics.* Chapman and Hall, London.

BISHOP, Y.M.M., FIENBERG, S.E. and HOLLAND, P.W. (1975). *Discrete Multivariate Analysis.* MIT Press, Cambridge, MA.

BLOOMFIELD, P. and WATSON, G.S. (1975). The inefficiency of least squares. *Biometrika* **62** 121–128.

BRADY, M.T., MCGRATH, N., BROUWERS, P., GELBER, R., FOWLER, M.G., YOGEV, R., HUTTON, N., BRYSON, Y.J., MITCHELL, C.D., FIKRIG, S., BORKOWSKY, W., JIMENEZ, E., MCSHERRY, G., RUBINSTEIN, A., WILFERT, C.M., MCINTOSH, K., ELKINS, M.M., WEINTRUB, P.S. and THE PEDIATRIC AIDS CLINICAL TRIALS GROUP (1996). Randomized study of the tolerance and efficacy of high- versus low-dose zidovudine in Human Immunodeficiency Virus-infected children with mild to moderate symptoms (AIDS Clinical Trials Group 128). *J. Infectious Diseases* **173** 1097–1106

BRESLOW, N.E. and CLAYTON, D.G.(1993). Approximate inference in generalized linear mixed models. *J. Amer. Statist. Assoc.* **88** 9–25.

BRESLOW, N.E. and LIN, X. (1995). Bias correction in generalized linear mixed models with a single component of dispersion *Biometrika* **82** 81–91.

CARROLL, R.J., WU, C.F.J. and RUPPERT, D. (1988). The effect of estimating weights in weighted least squares. *J. Amer. Statist. Assoc.* **83** 1045–1054.

COX, D.R. (1972). The analysis of multivariate binary data. *Appl. Statist.* **21**:113–20.

CRESSIE, N. (1991). *Spacial Statistics.* Wiley, New York.

CRESSIE, N. and LAHIRI, S.N. (1993). The Asymptotic Distribution of REML Estimators, *J. Multivariate Analysis* **45** 217–233.

DAWID, A.P. (1979). Conditional independence in statistical theory (with discussion). *J. Roy. Statist. Soc. Ser. B* **41** 1–31.

DE GRUTTOLA, V. and TU, X.M. (1994). Modeling progression of CD4-lymphocyte count and its relationship to survival time. *Biometrics* **50** 1003–1014.

DEMPSTER, A.P. LAIRD, N.M. and RUBIN, D.B. (1977). Maximum-likelihood with incomplete data via the EM algorithm. *J. Roy. Statist. Soc. Ser. B* **39** 1–38.

DIGGLE, P.J. (1988). An approach to the analysis of repeated measures. *Biometrics* **44** 959–971.

DIGGLE, P.J. and KENWARD, M.G. (1994). Informative drop-out in longitudinal data analysis. *Appl. Statist.* **43** 49–93.

DIGGLE, P.J., LIANG, K.Y. and ZEGER, S.L. (1994). *The Analysis of Longitudinal Data*, 1st ed. Clarendon Press, Oxford.

DOCKERY, D.W., WARE, J.H., FERRIS, B.G., JR, GLICKSBERG, D.S., FAY, M.E., SPIRO, A. and SPEIZER, F.E. (1985). Distribution of forced expiratory volume in one second and forced vital capacity in healthy white, adult never-smokers in six US cities. *Am. Rev. Respir. Dis.* **131** 511–20.

DONNELLY, C., LAIRD, N. M. and WARE, J.H. (1995). Predication and creation of smooth curves for temporarily correlated longitudinal data. *J. Amer. Statist. Assoc.* **90** 984–989.

DWYER, P.S. (1967). Some applications of matrix derivatives in multivariate analysis. *J. Amer. Statist. Assoc.* **62** 607–625.

EKHOLM, A. and SKINNER, C. (1998). The Muscatine children's obesity data reanalysed using pattern mixture models. *Appl. Statist.* **47** 251–263.

FISCHL, M.A., PARKER, C.B., PETTINELLI, C., WULFSOHN, M., HIRSCH, M.S., COLLIER, A.C., ANTONISKIS, D., HO, M., RICHMAN, D.D., FUCHS, E., MERIGAN, T.C., REICHMAN, R.C., GOLD, J., STEIGBIBEL, N., LEOUNG, G.S., RASHEED, S., TSIATIS, A. and THE AIDS CLINICAL TRIALS GROUP (1990). Randomized controlled trial of a reduced daily dose of Zidovudine in patients with the acquired immunodeficiency syndrome. *New England Journal of Medicine* **323** 1009–1019.

FISHER, R.A. (1925). Theory of statistical estimation. *Proc. Cambridge Phil. Soc.* **22** 700–725.

Fitzmaurice, G.M. and Laird, N.M. (1993). A likelihood-based method for analysing longitudinal binary responses. *Biometrika* **80** 141–51.

Fitzmaurice, G.M. and Laird, N.M. (2000). Generalized linear mixture models for handling nonignorable dropouts in longitudinal studies. *Biostatistics* **1** 141–156.

Fitzmaurice, G.M., Laird, N.M. and Lipsitz, S.R. (1994). Analysing incomplete longitudinal binary responses: a likelihood-based approach. *Biometrics* **50** 601–12.

Fitzmaurice, G.M., Laird, N.M. and Rotnitzky, A.G. (1993). Regression models for discrete longitudinal responses (with discussion). *Statist. Sci.* **8** 284–309.

Fitzmaurice, G.M., Laird, N.M. and Zahner, G.E.P. (1996). Multivariate logistic models for incomplete binary responses. *J. Amer. Statist. Assoc.* **91** 99–108.

Follman, D. and Wu, M. (1995). An approximate generalized linear model with random effects for informtive missing data. *Biometrics* **51** 151–168.

Freedman, D.A. and Peters, S. (1984). Bootstrapping a regression equation: some empirical results. *J. Amer. Statist. Assoc.* **79** 97–106.

Gilmore, A.R., Anderson, R.D. and Rae, A.L. (1985). The analysis of binomial data by a generalized linear mixed model. *Biometrika* **72** 593–599.

Glonek, G.F.V. (1999). On identifiability in models for incomplete binary data, *Statist. Probab. Lett.* **41** 191–197.

Glonek, G.F.V. and McCullagh, P. (1995). Multivariatae logistic models. *J. R. Stat. Soc. Ser. B Stat. Methodol.* **57** 533–546.

Goldstein, H. (1991). Nonlinear likelihood approaches to variance component estimation and to related problems. *J. Amer. Statist. Assoc.* **72** 320–340.

GREEN, P.J. (1987). Penalized likelihood for general semi-parametric regression models. *International Statistical Review* **55** 245–259.

GREENLEES, W.S., REECE, J.S. and ZIESCHANG, K.D. (1982). Imputation of missing values when the probability of response depends on the variable being imputed. *J. Amer. Statist. Assoc.* **77** 251–261.

GRIZZLE, J.E. and ALLEN, D.M. (1969). Analysis of growth and dose response curves. *Biometrics* **25** 357–381.

GRIZZLE, J.E., STARMER, C.F. and KOCH, G.G. (1969). Analysis of categorical data by linear models. *Biometrics* **25** 489–504.

HARVILLE, D.A. (1977). Maximum likelihood approaches to variance component estimation and to related problems. *J. Amer. Statist. Assoc.* **72** 320–338.

HARVILLE, D.A. (1999). *Matrix Algebra from a Statisticians Perspective.* Springer, New York.

HECKMAN, J. (1976). The common structure of statistical models of truncation, sample selection and limited dependent variables, and a simple estimator for such models. *Annals of Economic and Social Measurement* **5/4** 475–492.

HINDE, J. (1982). Compound Poisson regression models. In *GLIM 82: Proceedings of the International Conference on Generalized Linear Models* (R. Gilchrist, ed.) 109–121. Springer, New York.

HOGAN, J.W. and LAIRD, N.M. (1997a). Mixture models for the joint distribution of repeated measures and event times. *Statistics in Medicine* **16** 239–257.

HOGAN, J.W. and LAIRD, N.M. (1997b). Model-based approaches to analyzing incomplete longitudinal and failure time data. *Statistics in Medicine* **16** 259–272.

HOPPER, J.L., HIBBERT, M.E., MACASKILL, G.T., PHELAN, P.D. and LANDAU, L.I. (1991). Longitudinal analysis of lung function growth in healthy children and adolescents. *J. Appl. Physiol.* **70** 770–777.

HUBER, P. (1967). The behavior of maximum likelihood estimates under nonstandard conditions. In *Proceedings of the Fifth Berkeley Symposium on Mathematical Statistics and Probability* (L.M. LeCam and J. Neyman, eds.) 221–233. University of California Press, Berkeley.

HURNY, C., BERNHARD, J., COATES, A., CASTIGLIONE-GERTSCH, M., PETERSON, H., GELBER, R., FORBES, J., RUDESTAM, C., SIMONCINI, E., CRIVELLARI, D., GOLDHIRSCH, A. and SENN, H. (1996). Impact of adjuvant therapy on quality of life in women with node-positive operable breast cancer. *The Lancet* **347** 1279–1284.

JENNRICH, R.I. (1969). Asymptotic properties of nonlinear least squares estimators. *Ann. Math. Statist.* **20** 633–643.

JENNRICH, R.I. and SCHULCHTER, M.D. (1986). Unbalanced repeated-measures models with structured covariance matrices. *Biometrics* **42** 805–820.

JOHNSON, N.L. and KOTZ, S. (1970). *Distributions in Statistics, Continuous Univariate Distributions* **2**. Houghton-Mifflin, Boston.

JOHNSON, R.A. and WICHERN, D.W. (1992). *Applied Multivariate Analysis.* Prentice Hall, Englewood Cliffs, NJ.

KACKAR, R.N. and HARVILLE, D.A. (1981). Unbiasedness of two-stage estimation and prediction procedures for mixed linear models. *Comm. Statist. A–Theory Methods* **10** 1249–1261.

KALBFLEISCH, J.D. and SPROTT, D.A. (1970). Applications of likelihood methods to models involving large numbers of parameters (with discussion). *J. Roy. Statist. Soc. Ser. B* **32** 175–208. [Correction (1972) **34** 124–125].

KALBFLEISCH, J.D. and SPROTT, D.A. (1973). Marginal and conditional likelihoods, *Sankhyā Ser. A* **35** 311–328.

KALBFLEISCH, J.D. and SPROTT, D.A. (1976). On tests of significance (with discussion). In *Foundations of Probability Theory, Statistical Inference, and Statistical Theories of Science. Vol. II: Foundations and*

Philosophy of Epistemic Applications of Probability Theory 259–272. Reidel, Dordrecht.

KALBFLEISCH, J.D. and PRENTICE, R.L. (1980). *The Statistical Analysis of Failure Time Data.* Wiley, New York.

LAUER, R.M. (1975). Coronary heart disease risk factors in school children: The Muscatine study. *J. Pediatrics* **86** 697–706.

KNUIMAN, M. and LAIRD, N. (1990). Parameter estimation in variance component models for binary response data. In *Proceedings of the International Symposium on Advances in Statistical Methods for Genetic Improvement of Livestock* (D. Gianola and K. Hammond, eds.) 194–207. Springer, New York.

LAIRD, N.M. (1978). Empirical Bayes methods for two-way tables. *Biometrika* **65** 581–590.

LAIRD, N.M., DONNELLY, C. and WARE, J.H. (1992). Longitudinal studies with continuous responses. *Stat. Methods Med. Res.* **1** 3–25.

LAIRD, N.M., FITZMAURICE, G.M. and SCHWARTZ, A.G. (2000). The analysis of case-control data: epidemiologic studies of familial aggregation. *Handbook of Environmental and Public Health Statistics* **18** 465–482.

LAIRD, N.M. and WARE, J.H. (1982). Random effects models for longitudinal data: an overview of recent results. *Biometrics* **38** 963–974.

LANGE, N. and LAIRD, N.M. (1989). The effect of covariance structure on variance estimation in balanced growth-curve models with random parameters. *J. Amer. Statist. Assoc.* **84** 241–247.

LAPIERRE, Y.D., NAI, N.V., CHAUINARD, G., AWAD, A.G., SAXENA, B., JAMES, B., MCCLURE, D.J., BAKISH, D., MAX, P., MANCHANDA, R., BEAUDRY, P., BLOOM, D., ROTSTEIN, E., ANCILL, R., SANDOR, P., SLADEN-DEW, N., DURAND, C., CHANDRASENA, R., HORN, E., ELLIOT, D., DAS, M., RAVINDRA, A. and MATSOS, G. (1990). A controlled dose-ranging study of remoxipride and haloperidol in schizophrenia—a Canadian multicentre trial. *Acta Psychiatrica Scandinavia* **82** 72–76.

LIANG, K.Y. and ZEGER, S.L. (1986). Longitudinal data analysis using generalized linear models. *Biometrika* **73** 13–22.

LIANG, K.Y., ZEGER, S.L. and QAQISH, B. (1992). Multivariate regression analyses for categorical data (with discussion). *J. Roy. Statist. Soc. Ser. B* **54** 3–40.

LIN, X. and BRESLOW, N.E. (1996). Bias correction in generalized linear mixed models with multiple components of dispersion. *J. Amer. Statist. Assoc.* **91** 1007–1016.

LIPSITZ, S.R., LAIRD, N.M. and HARRINGTON, D.P. (1990). Maximum likelihood regression models for paired binary data. *Statistics in Medicine* **9** 1517–1525.

LIPSITZ, S.R., LAIRD, N.M. and HARRINGTON, D.P. (1991). Generalized estimating equations for correlated binary data using the odds ratio as a measure of association. *Biometrika* **78** 153–160.

LITTLE, J.A. (1995). Modeling the drop-out mechanism in repeated–measure studies. *J. Amer. Statist. Assoc.* **90** 1112–1121.

LITTLE, J.A. and RUBIN, D.B. (1987). *Statistical Analysis with Missing Data.* Wiley, New York.

LITTLE, R.J.A. (1993). Pattern-mixture models for multivariate incomplete data. *J. Amer. Statist. Assoc.* **88** 125–134.

LIU, C. and RUBIN, D.B. (1994). The ECME Algorithm: a simple extension of EM and ECM with faster monotone convergence. *Biometrika* **81** 633–648.

MACHIN, D., FARLEY, T., BUSCA, B., CAMPBELL, M. and D'ARCANGUES, C. (1988). Assessing changes in vaginal bleeding patterns in contracepting women. *Contraception* **38** 165–179.

MCCULLAGH, P. and NELDER, J.A. (1989). *Generalized Linear Models*, 2nd ed. Chapman and Hall, London.

MCCULLOCH, C.E. and SEARLE, S.R. (2001). *Generalized, Linear, and Mixed Models.* Wiley, New York.

McGilchrist, C.A. and Aisbett, C.W. (1991). Restricted BLUP for mixed linear models. *Biometrical J.* **33** 131–141.

McLachlan, G.J. and Krishnan, T. (1996). *The EM Algorithm and Extensions.* Wiley, New York.

Mori, M., Woodworth, G.G. and Woolson, R.F. (1992). Application of empirical Bayes inference of estimation of rate of change in the presence of informative right censoring. *Statistics in Medicine* **11** 621–631.

Morrison, D.F. (1990). *Multivariate Statistical Methods.* McGraw-Hill, New York.

Naumova, E.N., Must, A. and Laird, N.M. (2001). Evaluating the impact of "critical periods" in longitudinal studies of growth using piecewise mixed effects models. *International Journal of Epidemiology* **30** 1332–1341.

Neuhaus, J.M., Kalbfleisch, J.D. and Hauck, W.W. (1991). A comparison of cluster-specific and population-averaged approaches for analyzing correlated binary data. *Internat. Statist. Rev.* **59** 25–35.

Neyman J and Scott, E.L. (1951). On certain methods of estimating the linear structural relation between two variables. *Ann. Math. Statist.* **22** 352–361.

Newey, W.K. and McFadden, D. (1994). Large sample estimation and hypothesis testing. In *Handbook of Econometrics* (R.F. Engle and D.L. McFadden, eds.) **4** 2111–2245. North-Holland, Amsterdam.

Patterson, H.D. and Thompson, R. (1971). Recovery of inter-block information when block sizes are unequal. *Biometrika* **58** 545–554.

Prentice, R.L. (1988). Correlated binary regression with covariates specific to each binary observation. *Biometrics* **44** 1033–1048.

Rao, C.R. (1973). *Linear Statistical Inference and Its Applications*, 2nd ed. Chapter 4, p. 220. Wiley, New York.

REMON, M. (1984). On a concept of partial sufficiency: L-sufficiency. *Internat. Statist. Rev.* **52** 127–136.

ROBINS, J.M. (1987). A graphical approach to the identification and estimation of causal parameters in mortality studies with sustained exposure periods. *J. Chronic Diseases* **40** 139S–161S.

ROBINS, J.M. (1989). The control of confounding intermediate variables. *Statistics in Medicine* **8** 679–701.

ROBINS, J.M. and GILL, R.D. (1997). Non-response models for the analysis of non-monotone ignorable missing data. *Statistics in Medicine* **16** 39–56.

ROBINS, J.M., ROTNITZKY, A. and ZHAO, L.P. (1995). Analysis of semiparametric regression models for repeated outcomes in the presence of missing data. *J. Amer. Statist. Assoc.* **90** 106–121.

ROBINSON, G.K. (1991). That BLUP is a good thing: estimation of random effects. *Statist. Sci.* **6** 15–32.

RUBIN, D.B. (1976). Inference and missing data. *Biometrika* **63** 581–592.

RUBIN, D.B. (1978). Bayesian inference for causal effects: the role of randomization. *Ann. Statist.* **6** 34–58.

SCHALL, R. (1991). Estimation in generalized linear models with random effects. *Biometrika* **40** 917–927.

SCHULCHTER, M.D. (1992). Methods for the analysis of informatively censored longitudinal data. *Statistics in Medicine* **11** 1861–1870.

SCHWARTZ, A.G., YANG, P. and SWANSON, G.M. (1996). Familial Risk of lung cancer among nonsmokers and their relatives. *Amer. J. Epidemiology* **144** 554–562.

SEARLE, S.R. (1987). *Linear Models for Unbalanced Data.* Wiley, New York.

SEARLE, S.R., CASELLA, G. and MCCULLOCH, C.E. (1992). *Variance Components.* Wiley, New York.

SPROTT, D.A. (1975). Marginal and conditional sufficiency. *Biometrika* **75** 599–605.

STIRATELLI, R., LAIRD, N.M. and WARE, J.H. (1985). Random effects models for serial observations with dichotomous response. *Biometrics* **40** 961–972.

SZATROWSKI, T.H. (1980). Explicit maximum likelihood estimates from balanced data in the mixed model of the analysis of variance. *Ann. Statist.* **8** 811–819.

SZATROWSKI, T.H. and MILLER, J.J. (1980). Necessary and sufficient conditions for explicit solutions in the multivariate normal estimation problem for patterned means and covariances. *Ann. Statist.* **8** 802–810.

TROXEL, A.B., LIPSITZ, S.R. and HARRINGTON, D.P. (1998). Marginal models for the analysis of longitudinal measurements with nonignorable non-monotone missing data. *Biometrika* **85** 661–672.

TSIATIS, A.A., DE GRUTTOLA, V. and WULFSOHN, M.S. (1996). Modeling the relationship of survival to longitudinal data measured with error: applications to survival and CD4 counts in patients with AIDS. *J. Amer. Statist. Assoc.* **90** 27–37.

VONESH, E.F. and CARTER, R.L. (1992). Mixed effect nonlinear regression for unbalanced repeated measures. *Biometrics* **48** 1–18.

WALD, A. (1949). Note on the consistency of the maximum likelihood estimate. *Ann. Math. Statist.* **20** 595–601.

WARE, J.H., DOCKERY, D.W., LOUIS, T.A., XU, X., FERRIS, B.G. JR. and SPEIZER, F.E. (1990). Longitudinal and cross-sectional estimates of pulmonary function decline in never-smoking adults. *Amer. J. Epidemiology* **132** 685–700.

WEDDERBURN, R.W.M. (1974). Quasi-likelihood functions, generalized linear models, and the Gauss-Newton method. *Biometrika* **61** 439–447.

WHITE, H. (1980). A heteroskedasticity-consistent covariance matrix estimator and a direct test for heteroskedasticity. *Econometrika* **48** 817–838.

WOOLSON, R.F. and CLARKE, W.R. (1984). Analysis of categorical incomplete longitudinal data. *J. Roy. Statist. Soc. Ser. A* **147** 87–99.

WU, C.F. J. (1983). On the convergence properties of the EM algorithm. *Ann. Statist.* **11** 95-103.

WU, M.C. and BAILEY, K.R. (1988). Analysing changes in the presence of informative right censoring caused by death and withdrawal. *Statistics in Medicine* **7** 337–346.

WU, M.C. and BAILEY, K.R. (1989). Estimation and comparison of changes in the presence of informative right censoring: conditional linear model. *Biometrics* **45** 939–955.

WU, M.C. and CARROLL, R.J. (1988) Estimation and comparison of changes in the presence of informative right censoring by modeling the censoring process. *Biometrics* **44** 174–188. [Correction (1991) **47** 357.]

ZEGER, S.L. and LIANG, K.Y. (1986). Longitudinal data analysis for discrete continuous outcomes. *Biometrics* **42** 121–130.

ZEGER, S.L., LIANG, K.Y. and ALBERT, P.S. (1988). Models for longitudinal data: a generalized estimating equation approach. *Biometrics* **44** 1049–1060.

ZHAO, L.P. and PRENTICE, R.L. (1990). Correlated binary regression using a quadratic exponential model. *Biometrika* **77** 642–8.